Second Edition

RSM Simplified

Optimizing Processes Using Response Surface
Methods for Design of Experiments

Second Edition

RSM Simplified

Optimizing Processes Using Response Surface
Methods for Design of Experiments

Mark J. Anderson • Patrick J. Whitcomb

CRC Press
Taylor & Francis Group
Boca Raton   London   New York

CRC Press is an imprint of the
Taylor & Francis Group, an **Informa** business

A PRODUCTIVITY PRESS BOOK

CRC Press
Taylor & Francis Group
6000 Broken Sound Parkway NW, Suite 300
Boca Raton, FL 33487-2742

© 2017 by Taylor & Francis Group, LLC
CRC Press is an imprint of Taylor & Francis Group, an Informa business

Printed on acid-free paper
Version Date: 20160218

International Standard Book Number-13: 978-1-4987-4598-7 (Paperback)

Library of Congress Cataloging-in-Publication Data

Names: Anderson, Mark J., 1953- | Whitcomb, Patrick J., 1950-
Title: RSM simplified : optimizing processes using response surface methods for design of experiments / Mark J. Anderson and Patrick J. Whitcomb.
Other titles: Response surface methods simplified
Description: Second edition. | Boca Raton : Taylor & Francis, 2016. | "A CRC title." | Includes bibliographical references and index.
Identifiers: LCCN 2016003390 | ISBN 9781498745987 (alk. paper)
Subjects: LCSH: Experimental design. | Response surfaces (Statistics)
Classification: LCC QA279 .A52875 2016 | DDC 519.5/7--dc23
LC record available at http://lccn.loc.gov/2016003390

Visit the Taylor & Francis Web site at
http://www.taylorandfrancis.com

and the CRC Press Web site at
http://www.crcpress.com

Contents

Preface

Prediction is very hard, especially when it is about the future.

Yogi Berra

This book is a natural progression from our first book, *DOE Simplified: Practical Tools for Effective Experimentation*, originally published in 2000 and now in its third edition (2015). By concentrating on simple comparative and two-level factorial designs, *DOE Simplified* provides a topical overview of design of experiments (DOE). Although these relatively simple statistical tools for breakthrough may require just 20% of the overall experimental efforts, they likely provide 80% of the gain. From this you can infer that response surface methods (RSM)—the subject of this book—require the remaining 80% of effort but net only the final 20% of gain. This may not seem worthwhile, but in competitive situations in dog-eat-dog high-tech areas, a 20% edge makes all the difference for success versus failure.

RSM are a paradox in that although the graphical renderings in 3D are readily grasped by most people, even highly trained technical professionals find it daunting to get a complete understanding of the underlying principles of experimental design and mathematical modeling they're based on. However, the return on investment for education on RSM can be enormous, particularly for seemingly impossible missions to meet multiple response specifications on systems affected by a number of key factors.

Delivering RSM in simplified fashion presents a mighty challenge—on par with the slogan "Statistics Made Easy" that we try to live up to at Stat-Ease, Inc. It's a very fine line that Pat and I must walk between oversimplification and unnecessary detail. We make a good team in this regard in that I, with my education and experience being more on the business side, tend to gloss over the details and jump to the bottom line, while Pat happily dives into details on mathematics and statistics. When the two of us hang together, we form a stable balance as we tread this tightrope—at least we hope so: You, the reader, must be the judge.

Obviously our publisher would never have agreed to us writing about something as technical as RSM if the approach we used for *DOE Simplified* had not proved so successful. According to the principles of statistical process control (SPC), one should never tamper with success, so we will carry forward our template—short bursts of explanatory text and essential formulas, interspersed with sidebars offering amusements and statistical trivia.

Pat requested that we make *RSM Simplified* more academic, which I naturally resisted. However, I quickly realized that with such an advanced topic, it would be wrong to withhold all the academic details. In *RSM Simplified* you will see that some of the sidebars plus the appendices at the end of the chapters offer weightier mathematical material. These might be big boulders for nonstatistical types attempting to plow through the book, so we've taken them off the main track. If you're like me, you won't sweat these details, but those of you who, like Pat, thrive on math, will want to pore over every equation and perhaps work them out for yourselves. Knock yourself out! Along these lines, another departure from the format of *DOE Simplified* will be more liberal referencing to documents offering greater detail on technical material. Again, we hope to serve both types of readers, the experimenters who lack formal statistical education and the statisticians who must advise experimenters.

As we stated at the outset, *RSM Simplified* rests on the foundation we built with its predecessor—*DOE Simplified*. Therefore, to minimize redundancy, we must assume that you have achieved proficiency in the basics of design and analysis of experiments. In other words, *DOE Simplified* or its equivalent is a prerequisite for *RSM Simplified*. You will see numerous cross-references to *DOE Simplified* throughout this book in lieu of a rehash of material we've previously detailed.

A big factor in the success of *DOE Simplified* was its "hands-on" approach facilitated by inclusion of software tailored to the techniques taught in the book. It would be inconceivable in the twenty-first century to pursue RSM without the use of computers. We've made this very easy by providing as a companion to the book an advanced program with all the necessary tools to do RSM. Practice problems are provided at the end of each chapter, with directions on how to make use of the computational tools. If you're serious about gaining a working knowledge of RSM, we urge you to download the software, which is a fully functional, but time-limited educational version of a commercially available package. You will find details on how to do this in the "About the Software" page at the end of the book. There you are provided a path to a website offering the software download, plus files associated with the book. Throughout the book you will be directed to software tutorials that facilitate completion of practice problems. Even if you ultimately make use of different statistical software, you will round out your education on RSM by working through all the suggested tutorials. Also, try reproducing other cases cited in the text by opening and analyzing data files provided at the book's website. There you will also find solutions to problems not already detailed via software tutorials. A final heads-up: Check out the on-line help incorporated with the software—it offers a wealth of knowledge, not just about program features, but about the statistics themselves.

As we've discussed, *RSM Simplified* goes into a lot more detail than its predecessor—*DOE Simplified*. However, this time around we decided not to include all of the dozens of design templates developed for RSM. We can cover only the proverbial tip of the iceberg in the text. Within each major class of design, *RSM Simplified* provides at least one detailed design template, plus one or two practice problems. Beyond that, you must make use of the software to view a complete catalog of templates, or you can follow up on the references to original statistical articles specifying their construction. To illustrate the vast number of possible designs, in just one class of RSM—the "Box–Behnken" design (BBD)—the software program offers 19 unique templates allowing from 3 to 21 factors to be optimized. The largest Box-Behnken design requires 348 runs. Feel free to print this out from the software!

What's New in This Edition

A major new revision of the software that accompanies this book (via download from the Internet) sets the stage for introducing RSM designs where the randomization of one or more hard-to-change (HTC) factors can be restricted. These are called *split plots*—a structure to which we devoted an entirely new chapter (11) in the third edition of *DOE Simplified*. Given that foundation, less needs to be said about split plots in this book so it is incorporated in an additional chapter titled "Practical Aspects for RSM Success." There, in addition to providing insights on handling HTC factors via split plots, we detail a workaround to using power to "right-size" designs that makes use of a tool called "fraction of design space" (FDS). Also in this new chapter we lay out ways to confirm your RSM models, thus bringing the book to an appropriate conclusion: verifying (we hope!) a successful process-optimization experiment.

This edition adds a number of other developments in RSM, but, other than the new material on split plots, FDS, and confirmation, we kept it largely intact. Perhaps the biggest change with the second edition is it being set up in a format amenable to digital publishing. Now experimenters planet-wide can read *RSM Simplified*. All they need is an Internet connection.

Pat and I are indebted to the many contributors to development of RSM, but especially George Box. For many years we used his book written with Norm Draper, *Empirical Model-Building and Response Surfaces* (Wiley, 1987), as our bible for RSM. Stu Hunter, a coauthor with Box (and Bill Hunter) of the landmark text *Statistics for Experimenters* (Wiley, 1978), has been another source of inspiration for educating on tools for DOE. We also owe a debt of gratitude for many suggestions over the years from Doug Montgomery. His collaboration with Ray Myers and Christine Anderson-Cook on the *Response Surface Methodology* (Wiley, 2016) produced a top-shelf reference on this advanced topic. We make no attempt to compete at this educational level. Last, but not least, we acknowledge the statistical help provided for the first edition by our advisor Gary Oehlert and for the second edition by Martin Bezener.

Mark J. Anderson

Authors

Mark J. Anderson, PE, CQE, MBA, is a principal and general manager of Stat-Ease, Inc. (Minneapolis, Minnesota). A chemical engineer by profession, he also has a diverse array of experience in process development (earning a patent), quality assurance, marketing, purchasing, and general management. Prior to joining Stat-Ease, he spearheaded an award-winning quality improvement program (generating millions of dollars in profit for an international manufacturer) and served as general manager for a medical device manufacturer. His other achievements include an extensive portfolio of published articles on design of experiments (DOE). Anderson coauthored (with Whitcomb) *DOE Simplified: Practical Tools for Effective Experimentation*, 2nd Edition (Productivity Press, 2015).

Patrick J. Whitcomb, PE, MS, is the founding principal and president of Stat-Ease, Inc. Before starting his own business, he worked as a chemical engineer, quality assurance manager, and plant manager. Whitcomb developed Design-Ease® software, an easy-to-use program for design of two-level and general factorial experiments, and Design-Expert® software, an advanced user's program for response surface, mixture, and combined designs. He has provided consulting and training on the application of design of experiments (DOE) and other statistical methods for decades. In 2013, the Minnesota Federation of Engineering, Science and Technology Societies (MFESTS) awarded Whitcomb the Charles W. Britzius Distinguished Engineer Award for his lifetime achievements.

Chapter 1

Introduction to the Beauty of Response Surface Methods

> For a research worker, the unforgotten moments of his [or her] life
> are those rare ones, which come after years of plodding work,
> when the veil over nature's secret seems suddenly to lift, and when
> what was dark and chaotic appears in a clear and beautiful light
> and pattern.
>
> **Gerty Cori**
> *The first American woman to win a Nobel Prize in science*

Before we jump down to ground-level details on response surface methods (RSM), let's get a bird's-eye view of the lay of the land. First of all, we will assume that you have an interest in this topic from a practical perspective, not academic. A second big assumption is that you've mastered the simpler tools of design of experiments (DOE). (Don't worry, we will do some review in the next few chapters!)

RSM offer DOE tools that lead to peak process performance. RSM produce precise maps based on mathematical models. This methodology facilitates putting all your responses together via sophisticated optimization approaches, which ultimately lead to the discovery of sweet spots where you meet all specifications at minimal cost.

This answers the question: What's in it for me? Now let's see how RSM fits into the overall framework of DOE and learn some historical background.

Strategy of Experimentation: Role for RSM

The development of RSM began with the publication of a landmark article by Box and Wilson (1951) titled "On the Experimental Attainment of Optimum Conditions." In a retrospective of events leading up to this paper, Box (2000) recalled observing process improvement teams in the United Kingdom at Imperial Chemical Industries in the late 1940s. He and Wilson realized that, as a practical matter, statistical plans for experimentation must be very flexible and allow for a series of iterations.

Box and other industrial statisticians, notably Hunter (1958–1959) continued to hone the strategy of experimentation to the point where it became standard practice in chemical and other process industries in the United Kingdom and elsewhere. In the United States, Du Pont took the lead in making effective use of the tools of DOE, including RSM. Via their Management and Technology Center (sadly, now defunct), they took an in-house workshop called "Strategy of Experimentation" public and, over the last quarter of the twentieth century, trained legions of engineers, scientists, and quality professionals in these statistical methods for experimentation.

This now-proven strategy of experimentation, illustrated in Figure 1.1, begins with standard two-level fractional factorial design, mathematically described as "2^{k-p}" (Box and Hunter, 1961) or newer test plans with minimum runs (noted below), which provide a screening tool. During this phase, experimenters seek to discover the vital few factors that create statistically significant effects of practical importance for the goal of process improvement. To save time at this early stage where a number (k) of unknown factors must be quickly screened, the strategy calls for use of relatively low-resolution ("res") fractions (p).

A QUICK PRIMER ON NOTATION AND TERMINOLOGY FOR STANDARD SCREENING DESIGNS

Two-level DOEs work very well as screening tools. If performed properly, they can reveal the vital few factors that significantly affect your process. To save on costly runs, experimenters often perform only a fraction of all the possible combinations. There are many varieties of fractional two-level designs, such as Taguchi or Plackett–Burman, but we will restrict our discussion to the standard ones that statisticians symbolize as "2^{k-p}," where k refers to the number of factors and p is the degree of fractionation. Regardless of how you do it, cutting out runs reduces the ability of the design to resolve all possible effects, specifically the higher-order interactions. Minimal-run designs, such as seven factors in eight runs (2^{7-4})—a 1/16th (2^{-4}) fraction, can only estimate main effects. Statisticians label these low-quality designs as "resolution III" to indicate that main effects will be aliased with two-factor interactions (2FIs). Resolution III designs can produce significant improvements, but it's like kicking your PC (or slapping your laptop) to make it work: you won't discover what really caused the failure.

To help you grasp the concept of resolution, think of main effects as one factor and add this to the number of factors it will be aliased with. In resolution III, it's a 1-to-2 relation, which adds to 3. Resolution IV indicates a 1-to-3 aliasing ($1 + 3 = 4$). A resolution V design aliases main effects only with four factors ($1 + 4 = 5$).

Because of their ability to more clearly reveal main effects, resolution IV designs work much better than resolution III for screening purposes, but they still offer a large savings in experimental runs. For example, let's say that you want to screen 10 process factors ($k = 10$). A full two-level factorial requires 2^{10} (2^k) combinations, way too many (1024!) for a practical experiment. However, the catalog of standard two-level designs offers a 1/32nd fraction that's resolution IV, which will produce fairly clear estimates of main effects. To most efficiently describe this option mathematically, convert the fraction to 2^p ($p = 5$) scientific notation: 2^{-5} ($=1/2^5 = 1/(2 \times 2 \times 2 \times 2 \times 2) = 1/32$). This yields 2^{10-5} (2^{k-p}) and by simple arithmetic in the exponent ($10 - 5$): 2^5 runs. Now, we do the final calculation: $2 \times 2 \times 2 \times 2 \times 2$ equals 32 runs in the resolution IV fraction (vs. 1024 in the full factorial).

P.S. A more efficient type of fractional two-level factorial screening design has been developed (Anderson and Whitcomb, 2004). These designs are referred to as "minimum-run resolution IV" (MR4) because they require a minimal number of factor combinations (runs) to resolve main effects from 2FIs (resolution IV). They compare favorably to the classical alternatives on the basis of required experimental runs. For example, 10 factors can be screened in only 20 runs via the MR4 whereas the standard (2^{k-p}) resolution IV design, a 1/32nd (2^{-5}) fraction, requires 32 runs.

Along these same lines are minimum-run resolution V (MR5) designs, for example, one that characterizes six factors in only 22 runs versus 32 runs required by the standard test plan.

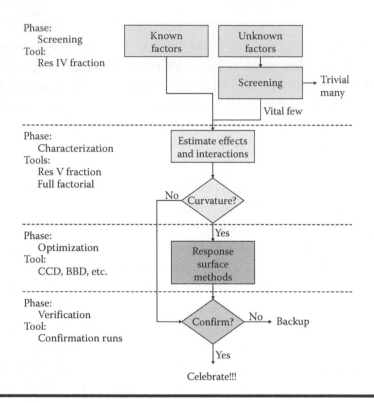

Figure 1.1   Strategy of experimentation.

After throwing the many trivial factors off to the side (preferably by holding them fixed or blocking them out), the experimental program should enter the characterization phase where interactions become evident. This requires higher-resolution, or possibly full, two-level factorial designs. By definition, traditional one-factor-at-a-time (OFAT) approaches will never uncover interactions of factors that often prove to be the key to success, so practitioners of statistical DOE often achieve breakthroughs at this phase. With process performance possibly nearing peak levels, center points (CPs) set at the mid-levels of the process factors had best be added to check for curvature.

As noted in the preface, 80% (or more) of all that can be gained in yield and quality from the process might be accomplished at this point, despite having invested only 20% of the overall experimental effort. However, high-tech industries facing severe competition cannot stop here. If curvature is detected in their systems, they must optimize their processes for the remaining 20% to be gained. As indicated in the flowchart in Figure 1.1, this is the point where RSM come into play. The typical tools used for RSM, which are detailed later in this book, are the central composite design (CCD) and Box–Behnken design (BBD), as well as computer-generated optimal designs.

After completing this three-part series of experiments according to the "SCO" (screening, characterization, and optimization) strategy, there remains one more step: verification. We will detail this final experimental stage in Chapter 12.

INTERACTIONS: HIDDEN GOLD—WHERE TO DIG THE HOLES? (A TRUE CONFESSION)

The dogma for good strategy of experimentation states that screening designs should *not* include factors known to be active in the system. My coauthor Pat, who seldom strays from standard statistical lines, kept saying this to students of Stat-Ease workshops, but I secretly thought he was out of his mind to take this approach. It seemed to me that it was like digging for gold in an area known to contain ore, but deliberately doing so in an area far from the mother lode. Finally, I confronted Pat about this and asked him to sketch out the flowchart just shown. Then it all made sense to me. It turns out that Pat neglected to mention that he planned to come back to the known factors and combine them with the vital few discovered by screening ones previously not known to have an impact on the system.

Mark

It was so dark, he [Stanley Yelnats] couldn't even see the end of his shovel. For all he knew he could be digging up gold and diamonds instead of dirt. He brought each shovelful close to his face to see if anything was there, before dumping it out of the hole.

***Holes* the Newberry Award-winning children's book by Louis Sachar (1998, p. 199), made into a popular movie by Walt Disney (2003)**

One Map Better than 1000 Statistics (Apologies to Confucius?)

The maps generated by RSM arguably provide the most value to experimenters. All the mathematics for the model fitting and statistics to validate it will be quickly forgotten the instant management and clients see three-dimensional (3D) renderings of the surface. Let's not resist this tendency to look at pretty pictures. Since we're only at the introductory stage, let's take a look at a variety of surfaces that typically emerge from RSM experiments.

PHONY QUOTATION

"A picture is worth one thousand words" has become a cliché that is invariably attributed to the ancient Chinese philosopher Confucius. Ironically, this oft-repeated quote is phony. It was fabricated by a marketing executive only about 100 years ago for ads appearing in streetcars.

Professor Paul Martin Lester's website
Department of Communications, Cal State University,
Fullerton, California

Life is really simple, but men insist on making it complicated.

Confucius

The first surface, done in wire mesh, shown in Figure 1.2 is the one you'd most like to see if increasing response is your goal.

Management and client would love to see this picture showing the peak of performance, but you likely will find it very difficult to attain, even with the powerful tools of RSM. More than likely you will hit operating boundaries well before reaching the peak. For example, the RPM on a machine might max out far short of where you'd like it for optimal response. In such cases, the result is likely to be a surface such as that

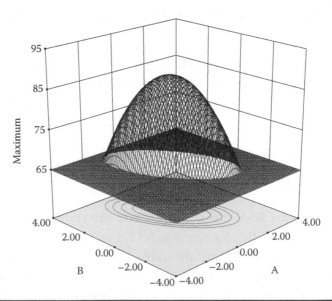

Figure 1.2   Simple maximum.

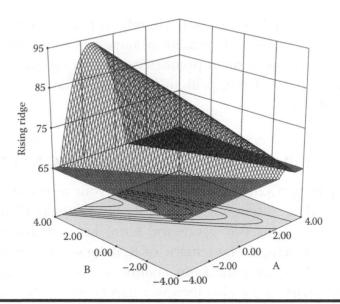

Figure 1.3   Rising ridge.

shown in Figure 1.3. This may be one of the most common geometries an experimenter will see with RSM.

Note the contours projected under the 3D surface. The individual contours represent points of constant response, much like a topographical map, shown as functions of two factors, in these cases A and B.

Figure 1.4 shows the most interesting surface that can be generated from standard RSM designs. This one is depicted with a solid exterior (as opposed to wire mesh). It exhibits a saddle, sometimes called a "mini-max."

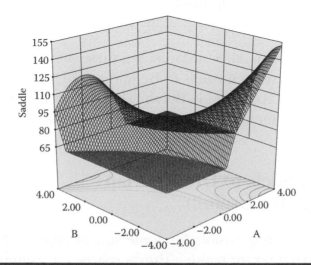

Figure 1.4   Saddle.

The model for a saddle varies only slightly from the simple maximum but the surfaces look quite different. More importantly, it presents major problems for finding a maximum point because one can get stuck in the saddle or push up to the lesser peak at left or right. The true optimum may get lost in the shuffle.

IMPRESS YOUR FRIENDS WITH GEOMETRIC KNOWLEDGE

To be strictly correct (and gain respect from your technical colleagues!) refer to this saddle surface as a "hyperbolic parabaloid." If you get a blank look when using this geometric term, just say that it looks like a Pringles® potato chip (a trademark of Kellogg's). This geometric form is also called a "col" after the Latin *collum* or neck collar. According to Webster's dictionary, a col also means a gap between mountain ranges. Meteorologists use the term col to mean the point of lowest or highest pressure between two lows or two highs. In the last century, daily newspapers typically published contour maps of barometric pressure in their section on weather, but it's hard to find these nowadays, perhaps because they look too technical.

If you really want to impress your friends with knowledge of geometry, drop this on them at your next social gathering: "monkey saddle." Seriously, this is a surface that mathematicians describe as one on "which a monkey can straddle with both his two legs and his tail." For a picture of this and the equations used to generate it, see http://mathworld.wolfram.com/MonkeySaddle.html.

We could show many other surfaces, but these are most common. What if you've got a response that needs to be minimized, rather than maximized? That's easy: turn this book upside-down. Then, for example, Figure 1.2 becomes a simple minimum. Now that's "RSM Simplified!"

Using Math Models to Draw a Map That's Wrong (but Useful!)

Assume that you are responsible for some sort of process. Most likely this will be a unit operation for making things (parts, for example) or stuff (such as food, pharmaceuticals, or chemicals) that can be manipulated via hands-on controls. This manufacturing scenario dominates the historical literature on RSM. However, as computer simulations become more and more sophisticated, it becomes useful to perform "virtual" experiments to save the time, trouble, and expense of building prototypes for jet engines and the like. The empirical predictive models produced from experiments on computer simulations are often referred to as "transfer functions." Another possible application of RSM could be something people related, such as a billing process for a medical institution, but few, if any, examples can be found for the use of this tool in these areas. The nature of the process does not matter, but to be amenable to optimization via RSM, it must be controllable by continuous, not categorical, factors. Obviously, it must also produce measurable responses, although even a simple rating scale like 1–5 will do.

It helps to consider your process as a black box system (see Figure 1.5) that gets subjected to various input factors, labeled "x." After manipulating the x's, the experimenter quantifies the responses by measurements y_1, y_2, etc.

For example, when baking a cake you can vary time (x_1) and temperature (x_2) to see how it impacts taste (y_1) and texture (y_2). This would be a fun experiment to do (one you should try at home!), but let's work on a simpler problem—when to get up in the morning and drive to work. Table 1.1 shows commuting times for Mark as a function of when he departed from home (Anderson, 2002). Rather than showing the actual times, it

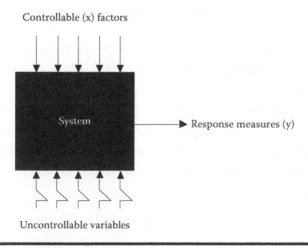

Figure 1.5 **System variables.**

Table 1.1 Commuting Times

x Departure (minutes)	y Drive Time (minutes)
0	30
2	38
7	40.4
13	38
20	40.4
20	37.2
33	36
40	37.2
40	38.8
47.3	53.2

turned out to be much more convenient to set time zero (t_0) at the earliest departure time of 6:30 a.m. Central United States. The table is sorted by departure time but the actual runs to work were done in random order to defray the impact of lurking time-related variables such as weather patterns and the like. Also note that some departure times were replicated to generate a measure of pure error. Finally, it must be confessed that Mark overslept one morning due to a random event involving his alarm clock, so the latest departure time was unplanned. He intended to depart at t_0, which would've provided a replicate, but it ended up being 47.3 minutes later. Such is life!

In the area where this study took place (Saint Paul and Minneapolis), most people work in shifts beginning at 7, 8, or 9 in the morning. Roads become clogged with traffic during a rush hour that typically falls between 7:00 a.m. and 9:00 a.m. With this in mind, it makes sense as a first approximation to try fitting a linear model to the data, which can be easily accomplished with any number of software packages featuring least-squares regression. The equation for actual factor levels, with coefficients rounded, is

$$\hat{y} = 34.7 + 0.19x$$

The hat (^) over the y symbolizes that we want to predict this value (the response of drive time).

CIRCUMFLEX: FRENCH HAT

Statisticians want to be sure you don't mix up actual responses with those that are predicted, so they put a hat over the latter. You will see many y's (guys?) wearing hats in this book because that's the ultimate objective for RSM—predicting responses as a function of the controllable factors (the x's). However, a friend of Mark recalled his statistics professor discussing putting "p" in a hat, which created some giggling by the class. To avoid such disrespect, you might refer to the hat by its proper name—the "circumflex." If you look for this in your word processor, it's likely to be with the French text, where it's used as an accent (also known as a "diacritic") mark. If you own Microsoft Word, look through the selection of letters included in their Arial Unicode MS font, which contains all of the characters, ideographs, and symbols defined in the Unicode standard (an encoding that enables almost all of the written languages in the world to be represented). Word provides a handy short-cut key for printing Unicode text. For example, to print the letter y with a circumflex ($\hat{y}$), type the code 0177 and then press Alt X. Voilà, you've made a prediction!

The line resulting from this predictive model is plotted against the raw data in Figure 1.6. (Note the axis for x being extended 10 minutes from the planned maximum of 40 minutes from the time-zero departure

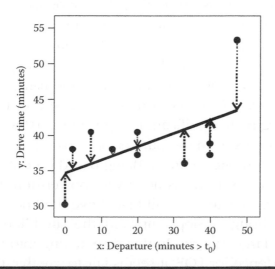

Figure 1.6   Linear fit of drive times.

Table 1.2 ANOVA for Drive-Time Experiment—Linear Model

Source	Sum of Squares (SS)	df	Mean Square (MS)	F-Value	p-Value Prob > F
Model	94.86	1	94.86	3.63	0.0933
Residual	209.16	8	26.14		
Lack of fit	*202.76*	*6*	*33.79*	*10.56*	*0.0890*
Pure error	*6.40*	*2*	*3.2*		
Cor Total	304.02	9			

time in order to cover for Mark oversleeping one of the days. If you set this problem up for computer analysis, enter the range as 0–50 rather than 0–40.)

The dashed, vertical lines represent the error or "residual" left over after the model fitting. The sum of squares (SS) of these residuals, 209.16, is a key element in the analysis of variance (ANOVA) shown in Table 1.2.

However, let's not get bogged down in the details of ANOVA. We went over this in *DOE Simplified*. An elegant mathematical explanation of ANOVA and other statistics applied to simple linear regression can be found in Weisberg (2013, Chapter 3).

What does Table 1.2 tell us about the significance of the model and how well it fits the data? First of all note the overall probability (Prob > F or p-value) of 0.0933, which might be interpreted as a marginally significant outcome. This outcome translates to 90.67% confidence (calculated by taking [1 minus the p-value] times 100) that the linear model coefficient is not zero. To keep it simple, let's just say we're more than 90% confident that there's an upslope to the drive-time data. Perhaps, we made some progress over simply taking the overall average of commuting times and using it for prediction.

That's the good news. Now the bad news: the test for lack of fit (LOF) also comes out marginally significant at a p-value of 0.089. Our benchmark for assessing LOF comes from the pure error pooled from the squared differences between the replicate times. Being based on only two degrees of freedom (df) for pure error, this is not a very powerful test, but notice in Figure 1.6 how the points tend to fall first above the line and then below. Furthermore, even though it slopes upward, the fitted line misses the highest time by a relatively large margin. In this case, taking into account the statistical and visual evidence for LOF, it seems fair to say that the linear model does *not* adequately represent the true response surface. Maybe it makes more sense to fit a curve to this data!

Neophytes to regression modeling often assess their success very simplistically with R^2. (See Appendix 1A entitled "Don't Let R^2 Fool You.") This statistic, referred to as the "coefficient of determination," measures the proportion of variation explained by the model relative to the mean (overall average of the response). It's easy to calculate from an ANOVA table:

$$R^2 = \frac{\text{Model SS}}{\text{Cor Total SS}}$$

The term "Cor Total" is statistical shorthand indicating that the SS is not calculated with zero as a basis, but instead it's been corrected ("Cor") for the mean.

For the drive-time case:

$$R^2 = \frac{94.86}{304.02} = 0.312$$

A special case occurs in simple linear models like the one we're evaluating for drive times: the coefficient of determination is simply the correlation coefficient (r) squared:

$$R^2 = r^2$$

The r-statistic ranges from −1, for a perfect *inverse* linear relationship, to +1 for a direct linear relationship. The correlation coefficient for drive time (y) versus departure (x) is +0.559, so

$$R^2 = r^2 = 0.559^2 = 0.312$$

Any way you calculate R^2, it represents less than one-third of the SS measured about the mean drive time, but it's better than nothing (0)! Or is it?

Consider what the objective is for applying RSM. According to the definition we presented at the beginning of the chapter, RSM are meant to lead you to the peak of process performance. From the word "lead," you can infer that the models produced from RSM must do more than explain what happened in the past—they must be of value for predicting the future. Obviously, only time will tell if your model works, but you could get tricky

(using "statis-tricks?") and hold back some of the data you've already collected and see if they are predicted accurately. Statisticians came up with a
great statistic that does something similar: it's called "predicted residual sum
of squares" or PRESS for short (Allen, 1971).

Here's the procedure for calculating PRESS:

1. Set aside an individual (i) observation from the original set (a sample
   of n points from your process) and refit your regression model to the
   remaining data (n − 1)
2. Measure the error in predicting point "i" and square this difference
3. Repeat for all n observations and compute the SS

Formulas for PRESS (Myers et al., 2016, Equation 2.50) have been
encoded in statistical software with regression tools, such as the program
accompanying this book. What's most important for you to remember
is that you want to minimize PRESS when evaluating alternative predictive models. Surprisingly, some models are so poor at prediction that their
PRESS exceeds the SS around the mean, called the "corrected total" (SS_{CorTot}).
In other words a "mean" model, which simply takes the average response
as the prediction, may work better than the model you've attempted to fit.
That does not sound promising, does it? A good way to be alerted when
this happens is to convert PRESS to an R^2-like statistic called "R-squared predicted" (R^2_{Pred}) by applying the following formula:

$$R^2_{Pred} = 1 - \left(\frac{PRESS}{SS_{CorTot}} \right)$$

You're probably thinking that R^2_{Pred} should vary from 0 (worst) to 1 (best)
like its cousin the raw R^2. No so—it often goes negative. That's not good!
Guess what happens when we compute this statistic for the linear model of
drive time?

$$R^2_{Pred} = 1 - \left(\frac{400.74}{304.02} \right) = -0.318$$

What should you do when R^2_{Pred} is less than zero? Maybe it will help to
illustrate PRESS graphically for a similar situation, but with fewer points.
Figure 1.7a and b illustrates bad PRESS for a fitting done on only four
response values.

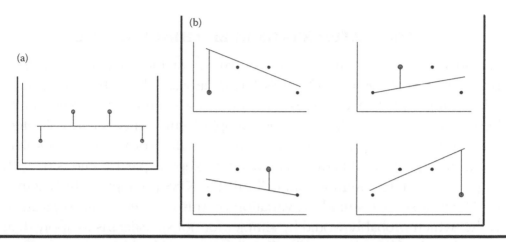

Figure 1.7 **(a) Residuals from mean. (b) PRESS (each point ignored).**

We hope you are not imPRESSed by this picture (sorry, could not resist the pun). Notice in Figure 1.7b that, as each response drops out (starting from the upper left graph), the linear model fits more poorly (deviations increase). Therefore, the PRESS surpasses the SS of the raw residuals (around the mean as reported in ANOVA) shown in Figure 1.7a. You can see that applying the linear model in this case does little good for prediction, so just taking the average, or "mean" model, would be better. It really would be mean to make Mark abandon all hope of predicting his drive time by excluding the connection with departure time embodied in the 0.19× term of the initially proposed model. Then, all he could do is use the overall average of approximately 39 minutes as a predictor, no matter what time he left home. That makes no sense given the utter reliability of traffic worsening over the daily rush hour of people commuting to work.

Mathematics comes to the rescue at this stage in the form of an approximation via an infinite series of powers of x published by the English mathematician Brook Taylor in the early 1700s, but actually used first by the Scotsman James Gregory (Thomas, 2000). For RSM, this function takes the form:

$$\hat{y} = \beta_0 + \beta_1 x_1 + \beta_{11} x_1^2 + \beta_{111} x_1^3 + \beta_{k\ldots k} x_1^k$$

The β symbols represent coefficients to be fitted via regression. It never pays to make things more complicated than necessary, so it's best to go one step at a time in this series, starting with the mean (β_0) and working up to the linear model ($\beta_0 + \beta_1 x_1$) and beyond.

HOW STATISTICIANS KEEP THINGS SIMPLE

Statisticians adhere to the principle of parsimony for model building. This principle, also known as Occam's Razor,[*] advises that when confronted with many equally accurate explanations of a scientific phenomenon it's best to choose the simplest one. For example, if sufficient data are collected, an experimenter might be tempted to fit the highest-order model possible, but a statistician would admonish that it be kept as parsimonious (frugal) as possible. More common folk call this the KISS principle, which stands for "keep it simple, stupid." A variation on this is "keep it simple, statistically," which probably would be simpler for nonstatisticians to apply than the principle of parsimony. The Boy Scouts provide a nice enhancement with their acronym KISMIF, which means "keep it simple and make it fun." Unfortunately, most people consider "statistics" and "simple" plus "fun" to be a contradiction in terms.

> I have only a high school education but I'm street smart, which can be more effective than college degrees. I operate under [the] rule … Keep it simple <u>and</u> stupid.

> **Jesse Ventura, former Governor of Minnesota, who made his name on the wrestling circuit as "The Body," but decided later to be known as "The Mind"**
>
> *November 1999 interview in* Playboy *magazine*

> Try to make things as simple as possible, but not simpler.
>
> **Einstein**

---

[*] Named for the medieval philosopher William of Occam.

Since we don't want to go back to the mean model for the driving data, but we know that the linear fit falls down on the job, the next step is to add a squared term to form the following second-order (also called "quadratic") model:

$$\hat{y} = 36.5 - 0.12x_1 + 0.0067x_1^2$$

Unfortunately, this higher-order model is not significant ($p \sim 0.2$), so it appears that we are making things worse, rather than better. Figure 1.8 does not contradict this view.

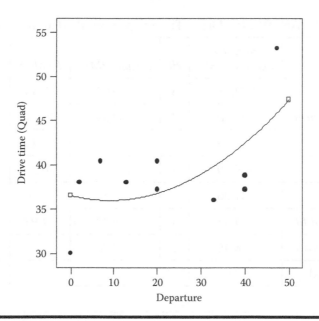

Figure 1.8   Quadratic fit of drive times.

The final nail in the quadratic coffin is the result for R^2_{Pred}, −1.177, which again is worse than nothing (just using the mean).

Let's not give up yet: try expanding the polynomial to the next level—add the third-order term:

$$\hat{y} = 32.1 + 1.7x_1 - 0.1x_1^2 + 0.0016x_1^3$$

According to the ANOVA in Table 1.3, this cubic polynomial does the trick for approximating the surface:

■ Highly significant overall model ($p < 0.01$ implying >99% confidence)
■ No significant LOF ($p \gg 0.1$)

Figure 1.9 shows the close agreement of this final model with the drive-time response.

The R^2_{Pred} is a healthy +0.632. The sharp-eyed statisticians seeing this plot will observe that the cubic model is highly leveraged by the highest drive time that occurred at the latest departure. (Recall that this was not part of Mark's original experimental plan.) If this one result proved inaccurate, the outcome would be quite different. In such case, the open square at the far right is an unwarranted extrapolation. Unfortunately, the probability of a morning rush hour is nearly on par with the sun coming up, but it's not documented in this particular dataset.

Table 1.3 ANOVA for Drive-Time Experiment—Cubic Model

Source	Sum of Squares (SS)	df	Mean Square (MS)	F-Value	p-Value Prob > F
Model	272.95	3	90.98	17.58	0.0022
A-Departure	59.60	1	59.60	11.51	0.0146
A^2	89.29	1	89.29	17.25	0.0060
A^3	160.20	1	160.20	30.94	0.0014
Residual	31.06	6	5.18		
Lack of fit	24.66	4	6.17	1.93	0.3696
Pure error	6.40	2	3.20		
Cor Total	304.02	9			

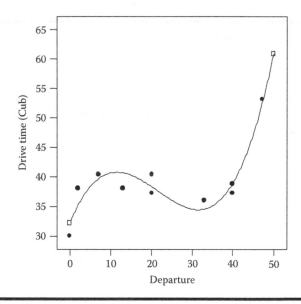

Figure 1.9 Cubic fit of drive times.

ASSESSING THE POTENTIAL INFLUENCE OF INPUT DATA VIA STATISTICAL LEVERAGE

Thanks to Mark's tardiness one morning we can learn something about how a model can be unduly influenced by one high-leverage point. concept of leverage is very easy to grasp when fitting a line to a collection of points.

To begin with, here's a give-away question: How many points do you need to fit a line? Of course the answer is two points, but what may not be obvious is that this relates to the number of coefficients in the linear model: one for the intercept (β_0) plus one for the slope (β_1). In general, you need at least one unique design point for every coefficient in the model.

Now for the next easy question: Would it be smart to fit a line with only two points? Obviously not! Each point will be fitted perfectly; no matter how far off it may be from reality. In this case, each of the points exhibits a "leverage" of 1. That's bad DOE! Leverage is easily reduced by simply replicating points. For example, by running two of each end-point the leverage of all four points drops to 0.5 and so on.

The general formula for leverage can be found in regression books such as that written by Weisberg (2013, p. 207). It is strictly a function of the input values and the chosen polynomial model. Therefore, leverages can and should be computed prior to doing a proper DOE. The average leverage is simply the number of model parameters (p) divided by the number (n) of design points. The cubic model for drive time contains four parameters (including the intercept β_0). The 10 runs then exhibit an average leverage of 0.4 (=4/10). However, not all leverages are equal. They range from 0.277 for the two departure times at 20 to 0.895 at the last departure of 47.3. As a rough rule of thumb, it's best to avoid relying on points with a leverage more than twice the average, which flags Mark for his tardy departure (0.895/0.4 > 2). If he had planned better for running late, Mark would've repeated this time of departure, thus cutting its leverage in half. His partner and coauthor Pat could hardly object to this reason for being tardy to work!

Another aspect of the fit illustrated in Figure 1.9 that would wither under a statistician's scrutiny is the waviness in the middle part that's characteristic of a cubic model. However, when pressed on this, Mark responds that the drive-time experiment represents a sample of 10 runs from literally thousands of commutes over nearly three decades. He insists on the basis of subject matter knowledge that times do not increase monotonically in the morning. According to him, there's a noticeable drop off for a time before the main rush. *Warning*: By taking Mark's expertise on his daily commute into account, we've illustrated biasing in model fitting which should be done only very judiciously, if ever.

Table 1.4 Model Summary Statistics for Drive-Time Experiment

Model	R^2 (Raw)	R^2_{Adj}	R^2_{Pred}	PRESS
Linear	0.312	0.226	−0.318	400.74
Quadratic	0.370	0.191	−1.177	661.83
Cubic	0.898	0.847	0.632	111.77

Table 1.4 provides summary statistics on the three orders of polynomials: first, second, and third, which translate to linear, quadratic, and cubic, respectively. This case may not stand up in court, but it makes a nice illustration of the mathematical aspects for model fitting in the context of RSM. As a practical matter, the results proved to be very useful for Mark. That is what really counts in the end.

ANOTHER ATTEMPT TO IMPROVE ON R^2

Notice that in Table 1.4, we list an adjusted version of R^2 (denoted by the subscript "Adj"). This adds little to the outcome of the drive-time experiment, so we mention this via a sidebar. It's actually a very useful statistic. However, because the R^2_{Pred} is even better suited for the purpose of RSM, we won't dwell on R^2_{Adj}. Here's our attempt to explain it in statistical terms and translate this to plainer language so you can see what's in it for you.

The problem is that the raw R^2 exhibits bias—a dirty word in statistics (with four letters, no less!). Bias is a systematic error in estimation of a population value. Who wants that? The bias in R^2 increases when naïve analysts cram a large number of predictors into a model based on a relatively small sample size. The R^2_{Adj} provides a more accurate (less biased) goodness-of-fit measure than the raw R^2.

What's in it for you is that the adjustment to R^2 counteracts the tendency for over-fitting data when doing regression. Recall from the equation shown earlier that the raw R^2 statistic is a ratio of the model SS divided by the corrected total SS ($R^2 = SS_{Model}/SS_{Cor\,Total}$). As more predictors come into the model, the numerator can only stay fixed or increase, but the denominator in this equation remains constant. Therefore, each additional variable you happen to come across can never make R^2 worse, only the same or larger, even when it's total nonsense.

For example, suppose that an erstwhile investor downloads all the data made available on the Internet by the US Department of Commerce's Bureau of Economic Analysis. It will then be very tempting to throw every possible leading indicator for stock market performance into the computer for regression modeling. It turns out that the raw R^2 will increase as each economic parameter is added to the model, even though it may produce nothing of value for response prediction. This would create a gross violation of the principle of parsimony.

Now, we get to the details. As you can see from the formula below, the adjustment to R^2 boils down to a penalty for model parameters (p):

$$R_{Adj}^2 = 1 - \frac{(n-1)}{(n-p)}(1-R^2)$$

Going back to the linear model for drive time we can substitute in 10 for n (the sample size) and 2 for p (the intercept and slope), plus the raw R^2:

$$R_{Adj}^2 = 1 - \frac{(10-1)}{(10-2)}(1-0.312) = 1 - \frac{9}{8}(0.688) = 1 - 0.774 = 0.226$$

As p, the number of model parameters increases, the penalty goes up, but if these parameters significantly improve the fit, the increase in raw R^2 will outweigh this reduction. If R_{Adj}^2 falls far below the raw R^2, there's a good chance you've included nonsignificant terms in the model, in which case it should be reduced to a more parsimonious level. In other words, keep it simple, statistically (KISS).

As you've seen in this example, the process of design and analysis with RSM is:

1. Design and execute an experiment to generate response data.
2. Fit the data to a series of polynomial models using tools of regression.
3. Conduct an ANOVA to assess statistical significance and test for LOF. Compare models on the basis of R_{Adj}^2 and R_{Pred}^2, but not raw R^2.
4. Choose the simplest model, even if only the mean, that predicts the response the best. Do not over-fit the data.

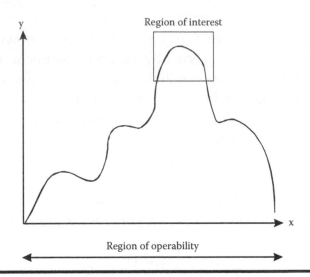

Figure 1.10   Region of interest versus region of operability.

Remember that with RSM you can only approximate the shape of the surface with a polynomial model. To keep things simple and affordable in terms of experimental runs, the more you focus your search the better. Use your subject matter knowledge and results from prior screening designs (see Figure 1.1) to select factors and ranges that frame a region of interest, within which you expect to find the ideal response levels. For example, as shown in Figure 1.10, it may be possible to operate over a very broad region of the input x. Perhaps by collecting enough data and applying a higher degree of polynomial, you could model the surface shown. But as a practical matter, all you really need is a model that works well within the smaller region of interest where performance (y) peaks.

If you achieve the proper focus, you will find that it often suffices to go only to the quadratic level (x to the power of 2) for your predictive models. For example, if you narrow the field of factors down to the vital two (call them A and B), and get them into the proper range, you might see simpler surfaces such as those displayed earlier in Figure 1.2, for example.

Now that we've described some of the math behind RSM, it may be interesting to view their underlying models:

- For the simple maximum (shown in Figure 1.2):
  $$\hat{y} = 83.57 + 9.39A + 7.12B - 7.44A^2 - 3.71B^2 - 5.80AB$$
- For the rising ridge (Figure 1.3):
  $$\hat{y} = 77.57 + 8.80A + 8.19B - 6.95A^2 - 2.07B^2 - 7.59AB$$

■ For the saddle (Figure 1.4):
$$\hat{y} = 84.29 + 11.06A + 4.05B - 6.46A^2 - 0.43B^2 - 9.38AB$$

Notice that these quadratic models for two factors include the 2FI term AB. This second-order term can be detected by high-resolution fractional or full two-level factorial designs, but the squared terms require a more sophisticated RSM design (to be discussed later in this book).

If you need cubic or higher orders of polynomials to adequately approximate your response surface, consider these alternatives:

■ Restricting the region of interest as discussed in the previous section
■ A transformation of the response via logarithm or some other mathematical function (see *DOE Simplified,* Chapter 4: "Dealing with Non-Normality via Response Transformations")
■ Looking for an outlier (for guidance, see Anderson and Whitcomb, 2003)

Ideally, by following up on one or more of these suggestions, you will find that a quadratic or lower order model (2FI or linear only) provides a good fit for your response data.

NO PRETENSES ABOUT ACCURACY
OF RSM FROM GEORGE BOX

George Box makes no apologies for the predictive models generated from RSM. He says:

All models are wrong. Some models are useful.

Only God knows the model.

Box and Bisgaard (1996)

Other insights offered by Box and Draper (1987, p. 413):

The exact functional relationship ... is usually unknown and possibly unknowable. ... We have only to think of the flight of a bird, the fall of a leaf, or the flow of water through a valve, to realize that, even using such equations, we are likely to be able to approximate only the main features of such a relationship.

Finding the Flats to Achieve Six Sigma Objectives for Robust Design

We have touted the value of mapping response surfaces for purposes of maximization, or on the flip side: minimization. Another goal may be to hit a target, such as product specifications, on a consistent basis. Decades ago, gurus such as Phil Crosby (1979) defined quality as "conformance to requirements," which allowed responses to range anywhere within the specified limits.

LONG AND SHORT OF "QUALITY"

"Quality" means the composite of material attributes including performance features and characteristics of a product or service to satisfy a given need.

US Department of Defense
Federal Acquisition Regulation 246.101

Quality is fitness for use.

J. M. Juran

In the early part of the twenty-first century, Six Sigma became the rage in industry. The goals grew much more ambitious: reduce process output variation so that six (or so—see sidebar) standard deviations (symbolized by the Greek letter sigma, σ) lie between the mean and the nearest specification limit. It's not good enough anymore to simply get product in "spec"—the probability of it ever going off-grade must be reduced to a minimum.

WHY SIX SIGMA? HOW MUCH IS ALLOWED FOR "WIGGLE-ROOM" AROUND THE TARGET?

The Six Sigma movement began at Motorola in the late 1980s. A joke made the rounds amongst statistical types that the slogan started as 5S but this did not go over with management. However, they found it hard to argue against 6S (to get the joke you must pronounce "6S" as "success"). Not only do we have a warped sense of humor, but also those of us in the know about statistics tend to get hung up on the widely quoted Six Sigma failure rate of 3.4 parts per million (ppm), or 99.99966% accuracy. According to the normal distribution the failure this far out in the tail should be much less—only two parts per billion. It turns out that Motorola allowed for some drift in the process mean, specifically 1.5 sigma in either direction (Pyzdek, 2003). The area of a normal distribution beyond 4.5 (=6–1.5) sigma does amount to 3.4 ppm. To illustrate what this level of performance entails, imagine playing 100 rounds of golf a year: at "six-sigma," even handicapped by the 1.5 sigma wiggle, you'd miss only 1 putt every 163 years!

The objective of Six Sigma is to find the flats—the high plateaus of product quality and process efficiency that do not get affected much by variations in factor settings. You can find these desirable operating regions by looking over the 3D renderings of response surfaces, or more precisely via a mathematical procedure called "propagation of error" (POE). We will get to the mathematics later in the book. For now, using the drive-time example for illustration, let's focus on the repercussions of varying factor levels on the resulting response.

From the work done earlier, Mark now feels comfortable that he can predict how long it will take him to get to work on average as a function of when he departs from home. Unfortunately, he cannot control the daily variations in traffic that occur due to weather and random accidents. Furthermore, the targeted departure time may vary somewhat, perhaps plus or minus 5 minutes. The impact of this variation in departure time (x) on the drive time (y) depends on the shape of the response surface at any particular point of departure. Figure 1.11 illustrates two of these points—labeled t_1 and t_2.

Notice the dramatic differences in transmitted variations y_1 and y_2. Obviously, it would be prudent for Mark to leave at the earlier time (mid-t_1) and not risk getting caught up in the rush hour delays at the upper end of t_2. The ideal locations for minimizing transmitted variation like this are obviously the flats in the curve. (There are two flats in the curve in Figure 1.11 but Mark prefers to sleep longer, so he chose the later one!) Finding these flats gets more difficult with increasing numbers of factors. At this stage, POE comes into play. However, we've got lots to do before getting to this advanced tool.

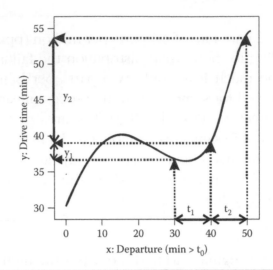

Figure 1.11 Variation in departure time for drive to work.

PRACTICE PROBLEM

1.1 The experiment on commuting time is one you really ought to try at home. Whether you follow up on this suggestion or do a different experiment, it will be helpful to make use of DOE software for

1. *Design.* Laying out the proper combinations of factors in random run order.

2. *Analysis.* Fitting the response data to an appropriate model, validating it statistically, and generating the surface plots.

For design and analysis of practice problems, and perhaps your own as well, we recommend you install the program made available to readers of this book. See the end of the book for instructions. To get started with the software, follow along with the tutorial entitled "*One-Factor RSM*" (* signifies other characters in the file names) which is broken into two parts—basic (required) and advanced (optional). This and other helpful user guides, posted with the program, contain many easy-to-follow screen shots; so, if possible, print them out to view side-by-side with your PC monitor. The files are created in portable document format, identified by the "pdf" extension. They can be read and printed with Adobe's free Acrobat® Reader program. The tutorial on one-factor RSM details how to set up and analyze the example on drive time. It includes instructions on how to modify a design for unplanned deviations in a factor setting. In Mark's case, he woke up 47.3 minutes late 1 day, so you must modify what the software specifies for that run.

Appendix 1A: Don't Let R^2 Fool You

Has a low R^2 ever disappointed you during the analysis of your experimental results? Is this really the kiss of death? Is all lost? Let's examine R^2 as it relates to DOE and find out.

Raw R^2 measures are calculated on the basis of the change in the response (Δy) relative to the total variation of the response ($\Delta y + \sigma$) over the range of the independent factor:

$$R^2 \cong \frac{\Delta y^2}{(\Delta y + \sigma)^2}$$

Let's look at an example where response y is dependent on factor x in a linear fashion:

$$y = \beta_0 + \beta_1 x$$

As illustrated by Figure 1A.1, assume a DOE is done on a noisy process (symbolized by the bell-shaped normal curve oriented to the response axis).

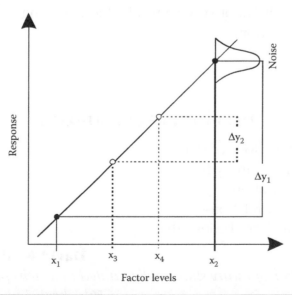

Figure 1A.1   Effect of varying factor levels (x) on response (y).

The experiment is done for purposes of estimating β_1—the slope in the linear model. Choosing independent factor levels at extremes of x_1 versus x_2 generates a large signal to noise ratio, thus making it fairly easy to estimate β_1. In this case, R^2 will be relatively high—perhaps approaching its theoretical maximum of one.

On the other hand, what if levels are tightened up to x_3 and x_4 as shown in Figure 1A.1? Obviously, this generates a smaller signal, thus making it more difficult to estimate the slope. This problem can be overcome by running replicates of each level, which reduces the variation of the averages. (This is predicted by the Central Limit Theorem as discussed in *DOE Simplified,* Chapter 1, "Basic Statistics for DOE.") If enough replicates are run, β_1 can be estimated with the same precision as in the first DOE, despite the narrower range of x_3 and x_4. However, because the signal (Δy) is smaller relative to the noise (σ), R^2 will be smaller, no matter how many replicates are run!

The goal of DOE is to identify the active factors and measure their effects. However, factor levels often must be restricted, even for experimental purposes. This is nearly always the case for studies done at the manufacturing stage, where upsets to the process cannot be risked. In such cases, success should be measured via ANOVA and (hopefully) its significant p-value, which can be achieved via the power of replication. When an experimenter succeeds on these measures, despite an accompanying low R^2, they should be congratulated for doing a proper job of DOE, not shot down for getting poor results! It boils down to this: although R^2 is a very popular and simple statistic, it is not very well suited to assessing outcomes from planned experimentation.

Don't be fooled by R^2!

DITTY ON R^2 CREDIBILITY

O, sing to the glory of stat!
Of sigma, x-bar and y-hat
The joy and elation
Of squared correlation—
Does anyone here believe that?

David K. Hildebrand
Author of many statistically slanted limericks published in
Hildebrand and Ott, 1998

Chapter 2

Lessons to Learn from Happenstance Regression

Experience does not ever err; it is only your judgment that
errs in promising itself results, which are not caused by your
experiments.

Leonardo da Vinci

As much as we'd like to begin building properly designed experiments
geared for RSM, it is necessary to first dispel the notion that you can save
time and money by simply gathering data from historical records. To some
extent, this chapter serves a similar purpose to the lectures given to teens
on the dangers of drugs, tobacco, sex, etc. Regression of happenstance data
may be every bit as addicting as these substances and activities. However,
we're all adults now, so a better reason for discussing happenstance regression may be to slip in some new statistics that can be extremely helpful for
diagnosing lousy layouts of input factors.

Learning by bad example is often much more interesting and therefore
more effective than displaying perfected demonstrations of technical virtuosity.
We hope this chapter generates thought on your part as to how better to
design RSM experiments and, more importantly, induce you to vow never to
try working around this by collecting historical data and running it through a
regression or neural network package.

Dangers of Happenstance Regression

The mathematical method of ordinary least squares, attributed by himself and others to Gauss (1809), set the stage for creating multilinear predictive models from historical data under the banner of "regression"—a term coined by Galton in 1885 (Stigler, 1986). These models are typically rated on the basis of R^2, which as you'll recall from the previous chapter, quantifies the explanation of variance on a scale of 0 (worst) to 1 (best). Unfortunately, when inputs become highly correlated—the usual condition for data collected at happenstance—the R^2 value becomes a poor indicator of a model's predictive value (Sergent et al., 1995).

For example, to test the accuracy of regression software, James Longley (1967) at the US Bureau of Labor Statistics looked at employment as a function of six factors:

A. Prices (GNP deflator, 1954 = 100)
B. GNP (gross national product)
C. Unemployment
D. Military (size of armed forces)
E. Population (people older than 14 years)
F. Time (year)

The raw data are shown in Table 2.1.

All of the input factors are interrelated so it's extremely difficult to unravel how each might affect the response—total employment. The high degree of correlation ($r > 0.99$) shown in Figure 2.1 for factors A versus B is not atypical.

Table 2.2 shows several ways that the employment response can be modeled as a function of the input factors identified by Longley. His data are sufficient only to fit the linear model. Higher-order models, even those containing interactions, become aliased. The first three models have been reduced by various methods (see note on "A Brief Word on Algorithmic Model Reduction").

A BRIEF WORD ON ALGORITHMIC MODEL REDUCTION

Techniques abound for reducing insignificant terms from regression models. Ideally, this would be done on data from a planned experiment by a subject matter expert. Then upon inspection of the ANOVA and statistical significance term-by-term, the reduction can be done intelligently. For example,

Table 2.1 Longley Data

Run #	A: Prices (1954 = 100)	B: GNP	C: Unemp.	D: Military (Armed Forces)	E: Pop. (People >14)	F: Time (Year)	Employ. Total
1	83	234,289	2356	1590	107,608	1947	60,323
2	88.5	259,426	2325	1456	108,632	1948	61,122
3	88.2	258,054	3682	1616	109,773	1949	60,171
4	89.5	284,599	3351	1650	110,929	1950	61,187
5	96.2	328,975	2099	3099	112,075	1951	63,221
6	98.1	346,999	1932	3594	113,270	1952	63,639
7	99	365,385	1870	3547	115,094	1953	64,989
8	100	363,112	3578	3350	116,219	1954	63,761
9	101.2	397,469	2904	3048	117,388	1955	66,019
10	104.6	419,180	2822	2857	118,734	1956	67,857
11	108.4	442,769	2936	2798	120,445	1957	68,169
12	110.8	444,546	4681	2637	121,950	1958	66,513
13	112.6	482,704	3813	2552	123,366	1959	68,655
14	114.2	502,601	3931	2514	125,368	1960	69,564
15	115.7	518,173	4806	2572	127,852	1961	69,331
16	116.9	554,894	4007	2827	130,081	1962	70,551

suppose that in a mechanical assembly, two factors, call them A and B, cannot physically interact and that the term modeling this does not turn out to be significant. It would make sense in this case to delete AB from the predictive model. Unfortunately, powerful computers and smart regression software make it very easy for the analyst to reduce models automatically via a variety of stepwise methods, such as

- ■ "Forward." Starting from the intercept, adding the next most significant term, and so on.
- ■ "Backward." Fit the highest-order model that will not contain aliased terms and then start eliminating the least-significant terms step-by-step.

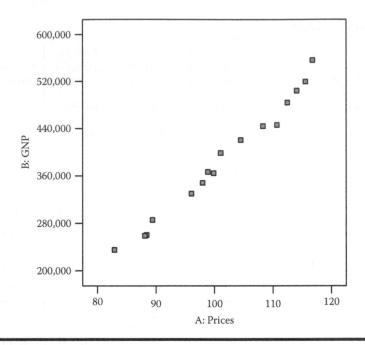

Figure 2.1 Correlation of factors in Longley dataset.

■ Combinations of the above such as first going forward and then going backward to drop terms that no longer meet significance criterion due to being replaced by better predictors.
■ Brute-force evaluation of all possible subsets (various combinations of terms). Choose the best by R^2_{Pred} or a similar model criterion.

This just gives you a rough idea in layperson's language of what can be done. For statistical details on how to accomplish model reduction, see Chapter 15 of Draper and Smith (1998).

Table 2.2 Model Coefficients (Rounded) for Longley Data

Model	β_0	A	B	C	D	E	F	R^2	R^2_{Pred}
1	52,400	NS	0.038	−0.54	NS	NS	NS	0.9807	0.9726
2	−3,600,000	NS	−0.040	−2.09	−1.01	NS	1890	0.9954	0.9892
3	−1,800,000	NS	NS	−1.47	−0.77	NS	956	0.9928	0.9885
4	−3,480,000			−2.02	−1.03		1830	0.9955	0.9844

Terms for which coefficients are not provided in Table 2.2 were not significant (NS) statistically using a p-value of 0.05 as the cutoff. The last model (number 4) includes all terms in the linear model, but those that are crossed out do not achieve statistical significance ($p > 0.05$).

Based on R^2, all three models evidently do a superb job of prediction. The second model comes out a bit ahead using the more discriminating R^2_{Pred} statistic. However, it's difficult to put much faith in any model given the huge discrepancy in the intercepts (β_0) and lack of commonality in the factors deemed significant versus NS.

At this point, it's helpful to introduce a very useful measure of input factor correlation called the "variance inflation factor" or "VIF" (Marquardt, 1970). Just as it says, this statistic quantifies the inflation in error for estimating model coefficients caused by the correlation of any individual (i) term with all the others. It's calculated as follows:

$$VIF = 1/(1 - R_i^2)$$

where R_i^2 is the multiple correlation coefficient of determination. An R_i^2 of zero is ideal, in which case it's considered in mathematical terms to be "orthogonal." Perhaps, you've heard the term "orthogonal matrix" used in an appreciative way by experts touting a particular DOE; for example, the classical two-level factorial (2^{k-p}) layouts. As you can see from Table 2.3 for model 4, the Longley data fall far short of orthogonality.

Notice that this table includes a computation of the standard error ("Std Err") of each term in the coded model. It's done on the basis of a unit (1) standard deviation for the response data. Thus, this information can be generated *before* running an experiment as part of a design evaluation.

Table 2.3 Nonorthogonality of Longley Data

Term	Std Err	VIF	R_i^2
A	4.72	136	0.993
B	17.6	1788	0.999
C	2.35	34	0.970
D	0.75	3.6	0.721
E	8.33	399	0.997
F	11.2	759	0.999

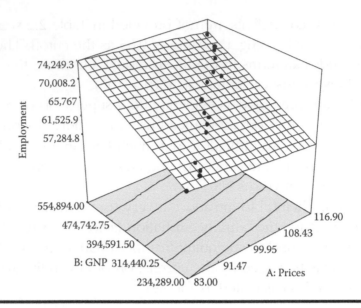

Figure 2.2 Longley data fitted to A and B.

You can see how the standard error goes up as the VIF increases. It actually goes up by the square root of the VIF. For example, if a coefficient has a VIF of 16, its standard error is four times as large as it would be in an orthogonal design.

The rule-of-thumb for VIF is that values of 10 or more indicate the associated regression coefficients are poorly estimated due to multicollinearity. Statisticians liken the pattern of two highly correlated inputs to a "picket fence" (Myers, 1986, p. 81). Figure 2.2 shows just the points of the picket fence for the first two factors in the Longley data.

Imagine trying to balance a sheet of plywood on the top of this fence!

The quest to forecast economies is quixotic—it will probably never succeed, but attempts will continue so long as there's money to be made. As Isaac Newton said in 1720 after losing his life savings in bad investments: "I can calculate the motions of heavenly bodies, but not the madness of people." Despite several centuries of developments in mathematics and computing power since Newton's time, attempts to model financial markets may be little more than a shot in the dark.

CONNECTION OF ECONOMICS TO STATISTICS

The German mathematician and political scientist Gottfried Achenwall (1719–1772) coined the term "statistics" to describe the summation of a nation's economic aspects. The word is a portmanteau of Latin and Italian words meaning "a description of state."

Joseph Sternberg, *Wall Street Journal, Book Review: The Leading Indicators*, by Zachary Karabell, February 18, 2014

Another lesson on the dangers of doing happenstance regression struck closer to home for the authors. Early in their careers as chemical engineers, they worked together in a process development group for a specialty chemical manufacturer. It was expected that studies on pilot plants be done according to sound statistical principles. However, once a process got established in the manufacturing plant it became very difficult to gain control to do a proper DOE. Invariably, the management suggested that engineers pull data from quality control (QC) records and run it through a regression analysis. They figured that by chance, any number of process factors would vary enough to more-or-less fill all the blanks in a DOE. No added expense would be incurred and there would be no risk of disrupting production. The authors recall an engineering colleague working full-time for months crunching numbers through a computer to produce apparently good statistical models. Unfortunately, models from any given month changed completely in terms of their input variables and none could predict responses well in any time period other than the one just past.

What if you are under pressure to produce a model from historical data and you won't be given the chance to validate it against future process performance? An idea that seems obvious once you've heard it is to hold back some of your data when doing the regression and see if they get predicted well. For example, let's say you've got access to 100 points of data: regress the first 90 and see how the model predicts the remaining 10. This is called "data splitting" (Snee, 1977). It requires that you partition your data into two subsamples:

■ Fitting sample
■ Validation sample

However, this technique works only if you've got more than enough data to produce a statistically significant model. As a rough rule of thumb, the data splitting approach should be used only if you can collect a sample size (n):

$$n \geq 2p + 20$$

where p is the number of model parameters. If you gather this much data (n or more samples), split it equally between fitting and validation samples.

Data splitting, as described above, often proves to be impractical or unpalatable (few feel comfortable fitting only a subset of data). A much more popular approach to data splitting is to apply the PRESS statistic as a measure of a model's predictive capability. We talked about PRESS in Chapter 1. It will be used throughout the book in the form of R^2_{Pred}, which as we discussed, is based on PRESS. However, it's hard to beat data splitting as an acid test of a model, other than actually waiting for the future to happen and then seeing how well it conforms to expectations.

GARBAGE-IN PRODUCES GARBAGE-OUT, BUT WHAT IF NO ONE KNOWS IT? (ANOTHER TRUE CONFESSION)

I remember when our process development group got its first pen plotter—attached to a primitive form of PC predating the first one produced by IBM. The computer could generate regression models and send them to the plotter for production of contour plots in multiple colors. Unfortunately, it was not smart enough to generate any statistics on the validity of the model. That did not prevent me from making a plot, which looked so pretty that I could not help but show it to my colleagues and management. Of course they all thought it was wonderful and never doubted its veracity. After all, it came from a computer! Needless to say, when I later subjected the model to statistical scrutiny it proved to be primarily due to random variations. Sadly, to this day, even technical people who really should know better often blindly take direction from mysterious electronic oracles, at least in the opinion of some (see quote below).

Mark

Engineers are quite comfortable these days—in fact, far too comfortable—with results from the blackest of black boxes: neural nets, genetic algorithms, data mining, and the like.

Russell Lenth
Statistics Professor, University of Iowa

On the other hand, by systematically making small changes to a process according to a statistically sound experimental plan, improvements are virtually guaranteed. For example, according to an expert in the field (DeVeaux, 2001), a client's DOE produced $15 million in additional profit in process improvements, but only after wasting many months on fruitless regression of historical data, done despite being warned that this would almost certainly produce nothing of value.

Why regression of happenstance data works so poorly:

1. The process is typically highly controlled so inputs and outputs vary little
2. Inputs tend to be highly correlated ($R^2 > 0.9$)

Why DOE works so well:

1. Broad changes can be made or small ones done repeatedly
2. Factors are controlled in ways that create little or no correlation ($R^2 \sim 0$)

PRACTICAL ACCUMULATED RECORD COMPARISONS ANALYSIS

The use of existing data to save on the effort of doing experimentation is sometimes referred to as practical accumulated record comparisons or "PARC." However, because such studies are retrospective, DOE expert Stu Hunter says that this acronym might apply better to planning after the research is completed. The value of data from such unplanned experiments can be described by spelling PARC backwards.

Richard DeVeaux
A guided tour of modern regression methods, Proceedings of the Section on Physical and Engineering Science, 1995, p. 10

The future ain't what it used to be.

Yogi Berra

The Storks-Bring-Babies Fallacy, in Which Correlation Implies Causation

Stories abound about how regression analysis proves that storks brought babies to Germany (or Copenhagen, etc.) after World War II (Sies, 1988). In some European countries, storks are considered a sign of good luck so

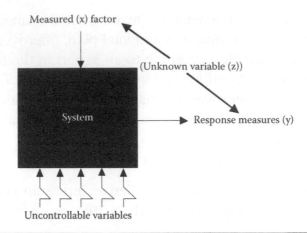

Figure 2.3 System variables.

it's wishful thinking that the high correlation ($R^2 \sim 0.96$) of these birds with babies implies causation. As population boomed after the war, new housing followed to accommodate it and storks nested in the chimneys. As illustrated in Figure 2.3, these spurious connections made by regression analysis stem from unknown variables.

As shown in the flow diagram, the x factor is not actually controlled, but only measured. It appears to create response y. However, in reality the unknown variable z, hidden from view of the analyst, affects both x and y. Now confusion reigns because the study, supported by regression models and accompanying statistics, reports a spurious cause-and-effect relationship between x and y.

DIET VERSUS HEART HEALTH

The Japanese eat very little fat and suffer fewer heart attacks than the British or Americans. The French eat a lot of fat and also suffer fewer heart attacks than the British or Americans. The Japanese drink very little red wine and suffer fewer heart attacks than the British or Americans. The Italians drink excessive amounts of red wine and also suffer fewer heart attacks than the British or Americans.

Conclusion: Eat and drink what you like. Speaking English is what kills you.

Irwin Knopf
Australian Readers Digest, November 2002, p. 73

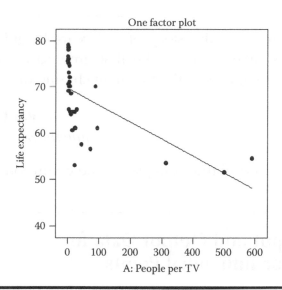

Figure 2.4   Effect of television on life expectancy.

Rossman (1994) provides an enlightening example of spurious correlation. He observed that life expectancy in various countries apparently varies with the number of people per television ("TV"). Figure 2.4 shows the regression line.

Notice that, as televisions become scarce (people per TV go up), the life expectancy declines. Could there be an unknown variable affecting both the factor (people per TV) and the response (life expectancy)? Should the richer countries ship boat-loads of televisions to the third world in order to improve their life expectancy? One wonders what kind of life it would be if it came at the cost of being forced to watch television!

DEER, OH DEAR!

In Minnesota, where the authors make their home, driving at dusk demands deer diligence due to the propensity of these woodland creatures to cross roads as the day ends. Bright yellow warning signs advise motorists that they are approaching deer-crossing areas. A child observing one of these signs asks, "Daddy, how do the deer know they should cross at the yellow signs?"

On a similar vein, the father takes his boy fishing for the first time and admonishes him to keep a close eye on the red and white bobber floating over the baited hook. The child asks, "How does the bobber know that it's got a fish?"

And, finally, there is the story of a 2-year-old girl who, going to her grandparents' lake cabin for the first time, saw a water-skier and exclaimed, "Grandma, look at the that man chasing the boat."

St. Paul Pioneer Press
Bulletin Board, June 14, 2010

These are three delightful examples of the confusion over correlation versus causation.

Diagnosing Input and Output Data for Outlying and/or Influential Points

The typical textbook examples are often too good to be true, so it's refreshing to see bad examples like this one on life expectancy versus televisions. In this case, we can learn how *not* to lay out an array of inputs: observe from Figure 2.4 that there a few extremely poor countries with many people per TV: from far right in descending order, these third-world countries are Myanmar (formerly Burma), Ethiopia, and Bangladesh. They exhibit an inordinate "leverage" on the slope of the predictive line. Small discrepancies in any one of these values would create big changes in the model. On the other hand, if any of the points at the left (representing rich countries such as the United States) changed, it would hardly matter—they exhibit low leverage. Figure 2.5 graphically illustrates the discrepancies in leverage.

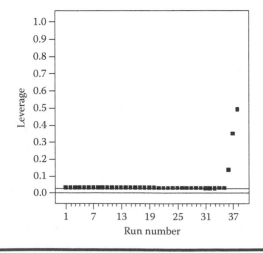

Figure 2.5 Leverage by country (run number).

The scale for leverage goes from 0 to 1. A leverage of 1 means that any error (experimental, measurement, etc.) associated with a response value will be carried into the model, so it will be predicted exactly. This measure of influence depends only on the input (x) values and the chosen model, so it can be determined by evaluating the DOE prior to conducting actual runs. High leverage points—those more than two times the average—can be "diluted" via replication. Two identical points will exhibit one-half the leverage of the original, three would drop it to one-third, and so forth. However, few people seem interested in replicating conditions in the third world so for better or worse we must deal with the data from Myanmar (formerly Burma), Ethiopia, and Bangladesh.

THE MATRIX RELOADED WITH X-FILES

If this were a movie, it would certainly appeal to statisticians! They thrive on matrices, for example, one constructed from the values of x, laid out run by run. For example, the statistical measure of leverage comes from the diagonal elements (h_{ij}) of a matrix **H** defined as follows:

$$\mathbf{H} = \mathbf{X}(\mathbf{X'X})^{-1}\mathbf{X'} = (h_{ij})$$

where the prime superscript (′) refers to the transpose and −1 indicates an inversion. (See Appendix 2A for an overview of matrix terminology and algebra.) This is commonly known as the "hat" matrix because it converts the vector of actual responses (**Y**) into predicted responses—the ones wearing the hats:

$$\hat{\mathbf{Y}} = \mathbf{HY}$$

We have no intention of getting into details on the matrix algebra that provides the superstructure for regression analysis. See Weisberg (2013, Appendix A.6, p. 278) for an elegant primer on this subject, or Montgomery et al. (2012, Appendix C.2, p. 577) for something with more meat. However, you might be interested to know that the hat matrix is symmetric and "idempotent" ($\mathbf{H}^2 = \mathbf{H}$). If you know someone who takes pride in their vast vocabulary, now you've got a new word to try on them. The true wordsmiths will puzzle out the mathematical meaning by knowing that "idem" is Latin for "the same" and potent comes from the Latin word "potens," which means able or powerful.

Let's get back to what's in the hat—the information about leverage. The average leverage equals the number of model coefficients, including the constant, divided by the number of runs in the experiment. A general guideline for leverage is that it should not exceed twice the average. For example, the average leverage (symbolized by h with a line on top) for the 38 countries that formed the basis for the linear model (two coefficients) on life expectancy versus people per TV is

$$\bar{h}_{ii} = 2/38 \approx 0.05$$

Thus, a value above 0.1 should be considered as an "outlier" in the independent variable space. The leverages for Myanmar, Ethiopia, and Bangladesh all exceed this threshold.

Have you ever had a dream, Neo, that you were so sure was real? What if you were unable to wake from that dream? How would you know the difference between the dream world and the real world?

The prophet Morpheus querying the hero of
***The Matrix* movie**

So far we've concentrated on the poor layout of input values for people per TV. What about the response data on life expectancy—do any of these appear to be outliers? Let's apply a tool called "deletion diagnostics" (Montgomery et al., 2012, p. 213). The idea is to measure the influence of each response after deleting it from the dataset. This requires refitting the model to what remains and looking for what differs with and without each individual.

The first deletion diagnostic we will discuss is called the "externally studentized residual," or more simply—the "R-student." The nonstatistician might find this alternative descriptor more understandable—"outlier t." Whatever you call it, this very useful statistic requires that each response be set aside, the model refitted and residual error calculated and, finally, plotted on a standard deviation scale. (Anticipating that this description will suffice, we will not expand on or lay out the formula for calculating externally studentized residuals, but you can find it in the reference by Weisberg.) A control chart of sorts can be produced by plotting this statistic versus the experimental runs, and then, superimposing "control" limits aimed at preventing tampering with the data. As a general rule, the upper and lower control limits should be placed at plus-or-minus 3.5 standard deviations. (More precise limits that take into account the size of experiment are provided by good DOE software.) Any individual runs that fall outside the limits should be investigated for special causes, such as typographical errors or mechanical breakdowns. If you establish that something abnormal occurred, ignore the outlying result and reanalyze the remainder of the response data. Results that fall within the control limits should be considered only common-cause variation. Removing any of this data would likely bias the outcome of your analysis.

GETTING SCHOOLED ON STUDENT

In the early 1900s, a chemist at Guinness brewery in Dublin, Ireland, developed tables showing the impact of sample size on variability. His name was W.S. Gosset, but he published under the pseudonym "Student." The term "studentized" is derived from Gosset and his legacy—the t-distribution, which corrects for the inherent error in estimating the true (population) standard deviation sigma (σ) in a normal distribution. Prior to Gosset, statisticians directed their attention to large samples of data, often with thousands of individuals (or they just ignored the fact that they estimated sigma). Their aims were primarily academic. Guinness couldn't afford to do pure research. He did only limited field trials on natural raw materials such as barley. Gossett/Student discovered that:

> As we decrease the number of experiments, the value of the standard deviation found from the sample of experiments becomes itself subject to increasing error.

> **W.S. Gosset**

Figure 2.6 shows the externally studentized residuals by country for the life expectancy as a function of people per TV.

Nothing stands out on this plot in regard to the upper and lower "control" limits. In most cases, one would take notice that the pattern is obviously not random, but here the data were sorted by factor level, so do not be alarmed.

Another deletion diagnostic to try on this case study is the DFFITS, which stands for difference in fits. DFFITS is calculated by measuring the change in each predicted value that occurs when that response is deleted. It is the

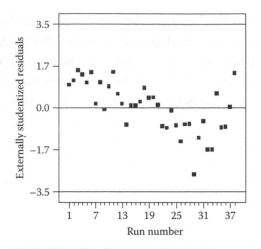

Figure 2.6   Externally studentized residuals (outlier t).

studentized difference between the predicted value with individual response i ($\hat{y}_{i,}$) and the predicted value without that individual ($\hat{y}_{i,-i}$). The equation is

$$\text{DFFITS}_i = \frac{\hat{y}_i - \hat{y}_{i,-i}}{s_{-i}\sqrt{h_{ii}}}$$

The larger the value of DFFITS the more it influences the fitted model. It can be shown mathematically (Myers, 1986, p. 284) that this statistic is the externally studentized residual magnified by high leverage points. This can be seen in Figure 2.7.

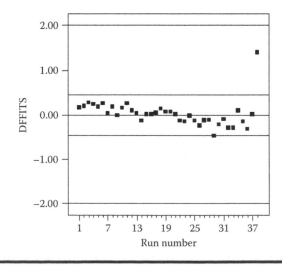

Figure 2.7   DFFITS by country.

Notice how the last point, representing the country of Myanmar, leaps above the rest. This results from it being high both in terms of leverage (Figure 2.5) and on the externally studentized residual plot (Figure 2.6). However, you'd better be careful before reacting to such a glaring discrepancy as this. It might be prudent to do some verification of data from Myanmar in this case, but there's really no reason to toss it out, because much of the discrepancy in DFFITS comes from this country being so starved for televisions relative to the rest of the world. It may be an outlier culturally and geographically, but not statistically. As Myers notes (1986, p. 84), "quality fit and quality prediction do not necessarily coincide."

OTHER DELETION DIAGNOSTICS

Yes, there are more statistics for assessing influence. If you are already overwhelmed by too much information (TMI), skip this sidebar!

The first statistic, closely related in concept to DFFITS, is called "DFBETAS." This deletion diagnostic breaks down how the deleted case changes each model coefficient, hence the acronym standing for difference in betas. For example, for one factor fit with a linear model to a given response, you will see two DFBETAS plots, one for the intercept and the other for the slope. The equation is

$$\text{DFBETAS}_{j,i} = \frac{\beta_j - \beta_{j,-i}}{s_{-i}\sqrt{c_{jj}}}$$

where i represents each individual run and j is the model coefficient. The quantity c_{jj} is the jth diagonal element of $(X'X)^{-1}$. Myers (1986, Section 8.3) gives cutoffs of $2/\sqrt{n}$ for DFBETAS and $2/\sqrt{p/n}$ for DFFITS. These can be easily calculated by regression software and placed on plots as guidelines for users. (The DOE software that accompanies this book places the DFBETAS and DFFITS limits at more conservative levels of $3/\sqrt{n}$ for DFBETAS and $3/\sqrt{p/n}$, respectively.)

Another measure of influence, essentially a distillation of DFBETAS, is Cook's distance (D), defined as

$$D_i = \frac{r_i^2}{p}\left(\frac{h_{ii}}{1 - h_{ii}}\right)$$

It is a product of the square of the internally studentized residual and a monotonic function of the leverage. Montgomery et al. (2012, p. 216) say

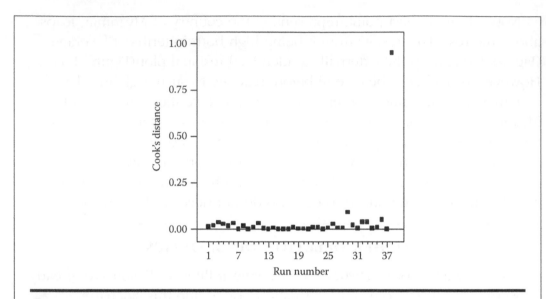

Figure 2.8   Cook's distance.

that the way to interpret Cook's is "the squared Euclidean distance…that the vector of fitted values moves when the ith observation is deleted." As a simple rule of thumb, they consider a Cook's distance greater than 1 to be influential. Rather than using 1 as a cutoff rule, we suggest all D_i values be examined via a plot versus run order. Look for gaps such as that shown in the plot for the case on life expectancy versus people per TV (see Figure 2.8).

Those poor people in Myanmar emerge again, same as on DFFITS, as being overly influential on the model. Maybe we should all donate our television sets. Who needs them (TV's) anyways?

…And I'm so worried about the shows on TV that sometimes they want to repeat.

Lyric from "I'm So Worried" song on Monty Python's
***Contractual Obligation* album**

Extrapolation Can Be Hazardous to Your Health!

As illustrated by the following example, if you stray outside the pocket of data you're likely to get sacked. The models created from regression cannot be relied upon outside the range of the factors tested. They become even more unreliable when applied to happenstance data.

American football fans spend an astounding amount of time poring over player statistics for gambling and fantasy gaming. Quarterback sacks are highly valued by these sports fans. Just for fun, we gathered statistics from the National Football League (NFL) on sacks for the 2002 season. The following demographics were collected for 167 defensive linemen who got at least one sack that year:

- Height ("H") in inches
- Weight ("W") in pounds
- Years ("Y") in NFL
- Games ("G") played
- Position (*E*nd, *T*ackle, or *N*ose)

For the defensive end position, which provides the clearest path to the opposing quarterback, regression on these factors produced the following, highly significant formula for predicting sacks:

$$\text{Log}_{10}\text{Sack}_E = -11 + 0.14\,H + 0.00032\,W + 1.3\,Y + 0.046\,G - 0.014\,H*Y - 0.00087\,W*Y$$

Note that response, sacks by ends (Sack_E), has been transformed by the base 10 logarithm to provide a better fit.

WHEN TACKLING TOUGH DATA, IT HELPS TO TAKE A LOG TO IT

Often as a response increases the standard deviation goes up in direct proportion. In other words, the error is not an absolute constant, as commonly assumed for statistical purposes, but rather it is a constant *percentage*. For example, a system may exhibit 10% error, so at a level of 10 the standard deviation is 1, but at a level of 100 the standard deviation becomes 10, and so on. Transforming such a response with the logarithm will stabilize the variance.

For details on this and alternative transformations, such as square root, see Chapter 4 in *DOE Simplified*.

Speak softly and carry a big stick; you will go far.

Theodore Roosevelt
Quoted as an African proverb in a 1901 speech

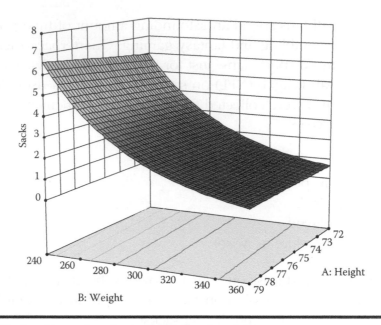

Figure 2.9   Defensive end sacks by height and weight.

As you can see by the response surface from the model shown in Figures 2.9, the lighter and taller players tend to get more sacks.

Using this equation to search out the ideal first-year (Y = 1) player, a numerical search revealed that the ideal player to draft for the defensive end position will have these physical attributes and predicted outcome: 84 inches (>2 meters!), 100 pounds (<50 kilograms!), producing an unheard of 60 sacks per year! These figures better describe an overgrown velociraptor or some other sort of genetically engineered predator, rather than a human being. Upon seeing such mad-sack creature, a quarterback might react like the tourists in the movie *Jurassic Park* seeing dinosaurs on the loose, "'Oooh! Ahhh!' That's how it always starts. Then later there's running and screaming" (spoken by the scientist Ian Malcolm).

Obviously, we allowed weight and height to be extrapolated outside of normal NFL bounds and thus created a nonsensical outcome. One tip-off that something's not right is that these two physical parameters exhibit nearly zero correlation (r = −0.08) within this exclusive class of professional defensive football players. Naturally, one would think that as height increases, so would weight. However, similar to not seeing the shape of a forest due to focusing too closely on the individual trees, known relationships of this sort become impossible to establish within highly restricted regions. This can be seen in Figure 2.10, which displays axes more inclusive of the general population. From this perspective, you can appreciate how

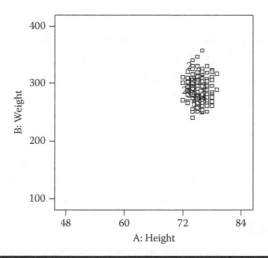

Figure 2.10   Scatter plot of height versus weight for NFL defensive linemen.

much the ranges of height and weight become skewed or distorted, both at the high end of their scales, once you focus on such an extremely small segment of the human species.

It can be very dangerous to put much faith in models derived from historical data like this where inputs fall within narrow limits, but cannot be independently controlled, and outputs vary on a happenstance basis.

FOOTBALL PARC ANALYSIS: HOW THE STOCK MARKET REACTS TO THE SUPER BOWL

Much has been made over the years about how US stock performance correlates to who wins the Super Bowl. The Super Bowl "theory" links US stock market performance to the results of the championship football game, held each January since 1967. It holds that if a team from the *original* NFL wins the title, the stock market increases for the rest of the year, but if a *new* team brought in from the *old* American Football League wins, the stock market goes down. Economist Paul M. Sommers of Middlebury College in Vermont debunked this theory by fitting a regression model to economic data for the years from 1967 to 1998. A much-simplified formula is

$$\text{Stock Index} = \beta_0 + \beta_1 \text{Original} + \beta_2 \text{New} + \varepsilon$$

For the first two decades of historical data, the formula works, explaining over 80% of the variation in stock by who wins the Super Bowl.

But just when investors became confident in football as a leading indicator, the correlation fell apart. For example, the Denver Broncos' championship in 1998 implied that stocks would go down more than 20%, but they actually went up over 20%.

Ivars Peterson
Super Bowls and stock markets,
Science News Online, July 1, 2000, Vol. 159, No. 1

PRACTICE PROBLEM

2.1 To see the extreme example of input correlation taken from the field of economic forecasting by Longley (1967), open the tutorial file titled "*Historical RSM*".pdf (* signifies other characters in the file names) posted with the computer program associated with the book. There you will learn how to do regression and ANOVA of happenstance data via the same software you were introduced to in Problem 1.1. You will learn how we produced the various models detailed in Table 2.2. See if you can do better!

Appendix 2A: A Brief Overview of Matrix Terminology

The X-matrix is a table consisting of n rows by p columns, where n represents the number of runs and p the parameters in the predictive model. Here's the X-matrix for the first attempt at modeling the drive-times as a function of departure time (from Chapter 1):

1	0
1	2
1	7
1	13
1	20
1	20
1	33
1	40
1	40
1	47.3

The model includes two parameters (p)—the intercept and the slope, so this matrix presents two columns, the first of which is all 1's and the other the actual input (x) values for departures.

The Y-matrix contains the actual response values, for example,

30
38
40.4
38
40.4
37.2
36
37.2
38.8
53.2

These are the times it took Mark to arrive at work in his morning commute.

To transpose a matrix, simply swap the rows and columns. For the X-matrix on departures shown above, this operation produces:

1	1	1	1	1	1	1	1	1	1
0	2	7	13	20	20	33	40	40	47.3

The transposed matrix is sometimes denoted with the same letter as the original matrix ($\mathbf{X}$) followed by a prime ($\mathbf{X'}$), but statisticians sometimes use the superscript T ($\mathbf{X}^T$) for clarity.

When multiplied by its parent $\mathbf{X}$, the *inverse* ($\mathbf{X}^{-1}$) of a matrix produces the identity ($\mathbf{I}$)—a square matrix with all ones on the main diagonal and zeroes elsewhere, such as

1	0	0
0	1	0
0	0	1

Algebraically, this is expressed as

$$XX^{-1} = X^{-1}X = I$$

Inverting matrices can be done by hand, but it becomes extremely tedious as p, the number of parameters, grows larger and the polynomial models increase in degree. At this stage, you run for the nearest computer loaded with software geared for matrix manipulation. If you want to see an example with the matrix math worked out on an RSM model (quadratic polynomial), see Box et al. (1978) Appendix 14B, pp. 501–502.

FISHER'S INFORMATION MATRIX

In the 1920s, R.A. Fisher invented the "information matrix"—**X'X**. (This should be shown divided by the variance (σ^2), but when designing experiments, σ^2 is fixed at 1.) The Fisher information matrix (FIM) plays a central role in statistical analysis and mathematical modeling of experimental data. It is instrumental for calculating confidence regions for model coefficients, and uncertainty bounds on the resulting predictions. The FIM also plays a key role in computer-generated "optimal" DOE as we will discuss later in the book.

> The million, million, million…to one chance happens once in a million, million…times no matter how surprised we may be that it results…
>
> **R.A. Fisher**

"What are the chances of a girl like you and a guy like me and me ending up together?"

"Not good."

"Not good like one in a hundred?"

"I'd say more like one in a million."

"So you're telling me there's a chance?"

Conversation in the movie *Dumb and Dumber* (1994) between the ever optimistic (but not too bright!) Lloyd, played by comedian Jim Carrey, and the glamorous Lauren Holly

Chapter 3

Factorials to Set the Stage for More Glamorous RSM Designs

When you come to a fork in the road, take it.

Yogi Berra

Ask just about anyone how an experiment should be conducted and they will tell you to change only one factor at a time (OFAT) while holding everything else constant. Look at any elementary science text and you will almost certainly find graphs with curves showing a response as a function of a single factor. Therefore, it should not be surprising that many top-notch technical professionals experiment via OFAT done at three or more levels.

Does OFAT really get the job done for optimization? Let's put this to the test via a simulation. Recall in Chapter 1 seeing a variety of response surfaces. We reproduced a common shape—a rising ridge—in Figure 3.1.

The predictive equation for this surface is

$$\hat{y} = 77.57 + 8.80A + 8.19B - 6.95A^2 - 2.07B^2 - 7.59AB$$

This will be our model for the following test of OFAT as an optimization method.

Putting OFAT in the Fire

Assume that the OFAT experimenter arbitrarily starts with factor A and varies it systematically over nine levels from low to high (−2 to +2) while

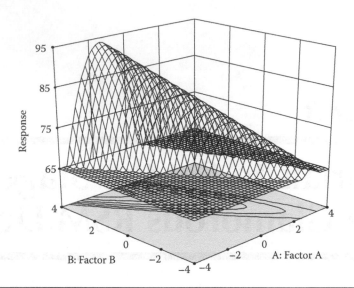

Figure 3.1   Rising ridge.

holding factor B at mid-level (0). The resulting response curve can be seen in Figure 3.2.

The experimenter sees that the response is maximized at an A-level near one (0.63 to be precise). The next step is then to vary B while holding factor A fixed at this "optimal" point. Figure 3.3 shows the results of the second OFAT experiment.

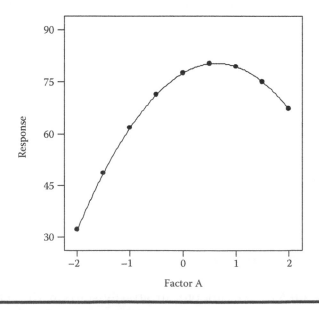

Figure 3.2   OFAT plot after experimenting on factor A.

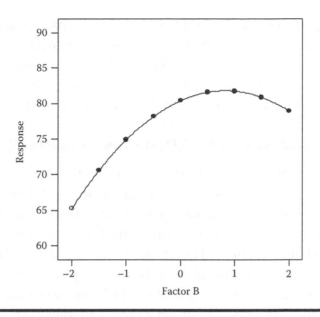

Figure 3.3   OFAT plot for factor B (A fixed at 0.63).

The response is now increased from 80 (previous maximum) to 82 by adjusting factor B to a level of 0.82 based on this second response plot. The OFAT experimenter proudly announces the optimum (A, B) combination of (0.63, 0.82), which produced a response above 80 in only 18 runs. However, as you can see in Figure 3.4, the real optimum is far higher (~94) than that attained via OFAT.

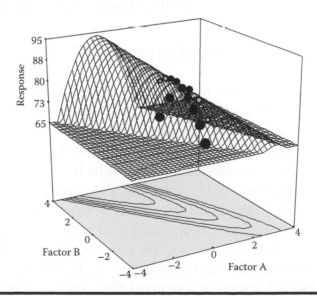

Figure 3.4   OFAT points shown on true surface.

By changing only one factor at a time, this experimenter got to the top of the ridge but did not realize that more could be gained by traveling up it. Furthermore, the results for the OFAT search on this surface will depend greatly on where B is fixed when A is varied and vice versa. A better DOE is needed to detect the interaction of AB that causes this unanticipated increase.

ADVICE FOR NINTH GRADERS ON EXPERIMENTATION

A science textbook used by ninth graders in the United States (Smith et al., 1993) advises that students vary only one condition while holding all others constant. They follow up with an activity that involves tying strings of varying length to two bolts, and hanging them from a string that also can differ in length. The students are asked to swing one of the bolts and record movement of the other (it's a physics thing!). Then it's suggested they change the weight of the bolt and see what happens. So, in the end, students experiment on many variables, but the follow-up question makes it clear how they should do it: "Why is it important to change only one variable at a time?" OFAT strikes again!

> The present method of experiment is blind and stupid; hence men wandering and roaming without any determined course, and consulting mere chance, are hurried about to various points, and advance but little…

> **Francis Bacon**
> *Novum Organum,* First Book, 70 (1620)

Factorial Designs as an Option to OFAT

To put OFAT to rest as an effective approach for optimization, what would be a good alternative? The obvious solution is a three-level factorial design—symbolized mathematically as 3^k, where k represents the number of factors. For example, in the case just discussed, having two factors, the 3^k design produces nine unique combinations ($3^2 = 9$), which exceeds the number of coefficients in the quadratic polynomial (six including the intercept). Thus, a three-level design would be a far better choice of design than OFAT for RSM.

This seems simple enough: to optimize your process run all factors at three levels. However, as k increases, the number of runs goes up exponentially relative to the number needed to estimate all parameters in the quadratic model. Table 3.1 shows that beyond three factors, the number of runs becomes excessive.

Table 3.1 Runs for Three-Level Factorials versus Coefficients in a Quadratic Model

Factors (k)	3^k Runs	Quadratic Coeff.
2	9	6
3	27	10
4	81	15
5	243	21

The number of runs required for 3^k becomes really ridiculous beyond three or four factors. It's conceivable that a linear model might do just as well as a quadratic for predicting response within the ranges of factors as tested. In this case, you'd not only waste many runs (and all the time and resources), but also do all the work of setting three levels when two would do. That would appear very foolish after the fact. We really ought to go back a step to two-level factorials (2^k) and see what can be done with these simpler designs as a basis for RSM.

**WHEN YOU GET A NEW HAMMER,
EVERYTHING LOOKS LIKE A NAIL**

A client of Stat-Ease had brought in another consultant who specialized in three-level designs for RSM. The designs were not necessarily full 3^k, but they provided a sufficient number of runs to fit a quadratic polynomial. The good news is that this client abandoned OFAT in favor of RSM. The bad news is that they used RSM for every problem, even when it was likely that conditions were far from optimum. This might be likened to pounding a finishing nail with a sledge hammer. A great deal of energy was wasted due to over-designing experiments until this client learned the strategy of first applying two-level fractional factorials (2^{k-p}) for screening, followed up by high-resolution designs with CPs, and finally, only if needed, RSM.

Two-Level Factorial Design with CPs: A Foundation for RSM

Let's assume that you've embarked on a process improvement project using the strategy of experimentation map shown in Figure 1.1 (Chapter 1). We pick up the story part-way down the road at the stage where you've completed screening studies and want to focus on a handful of factors known to be active in your system. For example, a chemist wanting to optimize yields narrows the field down to three key inputs with the ranges shown below:

A. Time, minutes: 80–100
B. Temperature, degrees Celsius: 140–150
C. Rate of addition, milliliters/minute: 4–6

This experimenter is certain that interactions exist among these factors, but perhaps no effects of any higher order will occur. This scenario fits the two-factor interaction "2FI" model:

$$\hat{y} = \beta_0 + \beta_1 A + \beta_2 B + \beta_3 C + \beta_{12} AB + \beta_{13} AC + \beta_{23} BC$$

This model does not require a full 3^k design, a 2^k will do. However, to be safe, it's best in cases like this to add CPs, replicated at least four times to provide sufficient information on possible curvature in the system. Figure 3.5 shows a picture of this design for three factors ($k = 3$).

The 2^k design (cubical portion) produces eight unique combinations, enough to fit the 2FI model, which contains only seven coefficients. The savings of doing only the 2^3 versus the full 3^3 is obviously substantial. However, you get what you pay for: the chosen design will not do much better than OFAT at revealing the sharp curvature shown in the rising ridge in Figure 3.1. That's where CPs come into play. As we will explain further, although they provide no information for estimating any of the 2FI coefficients, CPs support a test for curvature in the form of pure quadratic terms (A^2, B^2, C^2).

Table 3.2 shows the result of the experiment by the chemist. Note that yield is measured as grams of product made as a result of the reaction. The data are listed in a standard order used for design templates, but the experiment was actually done by randomizing the runs.

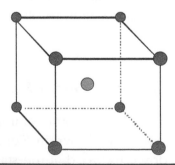

Figure 3.5   Two-level factorial design with CP.

Table 3.2   Two-Level Factorial with CPs on Chemical Reaction

Std Ord.	A: Time (minutes)	B: Temp. (degree Celcius)	C: Rate (mL/minutes)	Yield (grams)
1	80 (–)	140 (–)	4 (–)	76.6
2	100 (+)	140 (–)	4 (–)	82.5
3	80 (–)	150 (+)	4 (–)	86.0
4	100 (+)	150 (+)	4 (–)	75.9
5	80 (–)	140 (–)	6 (+)	79.1
6	100 (+)	140 (–)	6 (+)	82.1
7	80 (–)	150 (+)	6 (+)	88.2
8	100 (+)	150 (+)	6 (+)	79.0
9	90 (0)	145 (0)	5 (0)	87.1
10	90 (0)	145 (0)	5 (0)	85.7
11	90 (0)	145 (0)	5 (0)	87.8
12	90 (0)	145 (0)	5 (0)	84.2

OBTAINING A TRULY "PURE" ERROR

When viewing the layout of factors in Table 3.2, it's tempting to actually execute the experiment in the order (standard) shown, especially when it comes to the CPs (listed in rows 9–12). This would be a big mistake, particularly in a continuous (as opposed to batch) reaction, because the variation would be far less this way than if the four runs were done at random intervals throughout the experiment. A proper replication would require that the same steps be taken as for any other run; that is, starting by charging the reactor, then bringing it up to temperature, etc. In our consultancy at Stat-Ease, we've come across innumerable examples where so-called "pure error" came out suspiciously low. After questioning our clients on this, we've found many causes for incorrect estimates of error (listed in order of grievousness):

■ Running replicates one after the other
■ Rerunning replicates that don't come out as expected (common when factors are set up according to standard operating procedures)

> ■ Taking multiple samples from a given run and terming them "replicates"
> ■ Simply retesting a given sample a number of times
> ■ Recording the same number over and over
>
> You may scoff at this last misbehavior (known in the chemical industry as "dry-labbing") but this happened to a Stat-Ease client—a chemist who laid out a well-constructed RSM design to be run in random order by a technician. The ANOVA revealed a pure error of zero. It was overlooked until we pointed out a statistic called "lack of fit" (discussed somewhat in Chapter 1 and in more detail in this chapter), which put the estimate of pure error in the divisor, thus causing the result to be infinite. Upon inspection of the data, we realized that the technician wrote down the same result to many decimal places, so obviously it was thought to be a waste of time rerunning the reaction.

Notice that in Table 3.2, we've put the coded levels of each factor in parentheses. The classical (we prefer to call it "standard") approach to two-level design converts factors to –1 and +1 for low versus high levels, respectively. This facilitates model fitting and coefficient interpretation.

Predictive equations are generally done in coded form so they can be readily interpreted. In this case, the standard method for analysis of 2^k designs (see *DOE Simplified*, Chapter 3) produced the coded equation shown below

$$\text{Yield} = 82.85 - 1.30A + 1.10B + 0.92C - 3.52AB$$

The value for the intercept ($\beta_0 = 82.85$) represents the average of all actual responses. In coded format, units of measure for the predictors (A, B, and C in this case) become superfluous. Therefore, the model coefficients (β) can be directly compared as a measure of their relative impact. For example, in the equation for yield, note the relatively large interaction between factors A and B and the fact that it's negative. This indicates an appreciable antagonism between time and temperature. In other words, when both these factors are at the same level, the response drops off. (They don't get along well together!) The interaction plot for AB is displayed in Figure 3.6.

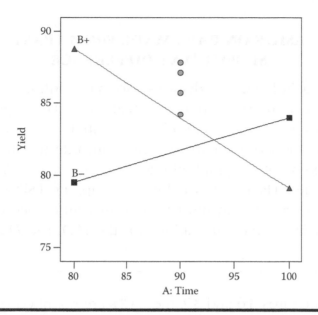

Figure 3.6   Interaction plot for effect of AB on reaction yield.

The points in the middle represent results from the four CPs, the values of which can be seen in the last four rows of Table 3.2: 87.1, 85.7, 87.8, and 84.2. The CPs remain at the same levels on the main effect plot (Figure 3.7) for factor C. We included this factor in the model only for illustrative purposes. It was not significant ($p > 0.1$).

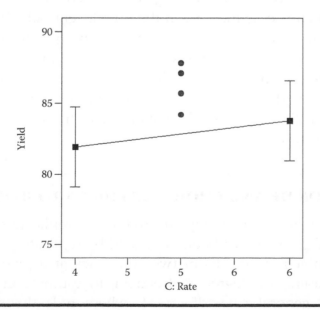

Figure 3.7   Main effect plot for factor C.

**ASIDE ON BARS MADE FROM LEAST
SIGNIFICANT DIFFERENCE**

The effects plot in Figure 3.7 displays a form of confidence interval (CI) called a "least significant difference" (LSD) bar. These particular LSD bars are set to provide 95% confidence in differences. In other words, when the bars do not overlap, you can conclude that the differences in means (end points of graphed lines) are significant. The line showing the C's effect (Figure 3.7) exhibit overlapping LSD bars from left to right, so the conclusion must be that this difference is *not* significant. For details on the calculations for the LSD, see *DOE Simplified,* Chapter 2.

Clearly, from Figures 3.6 and 3.7, these CPs are not fitted very well. They appear to be too high. According to the standard coding scheme (–/+), the CP falls at zero for all factors ([A,B,C] = [0,0,0]). When you plug these coordinates into the coded equation, the result equals the intercept: 82.5. This predicted response for the CP falls below *all* of the actual values. Something clearly is not right with the factorial model for the reaction yield. It fails to adequately fit what obviously is a curve in the response.

At this stage, we can take advantage of having properly replicated (we hope: see the note titled "More on Degrees of Freedom...") one or more design points, in this case the ones at the overall centroid of the cubical region of experimentation. (Sorry, but to keep you on your toes, we got a bit geometric there!) The four CP responses produce three measures, statistically termed "degrees of freedom" (df), of variation known as "pure error." You may infer from this terminology the existence of "impure" error, which is not far off—another estimate of variation comes from the residual left over from the model fitting.

MORE ON DF AND CHOICE OF POINTS TO REPLICATE

In the reactor case, it may appear arbitrary to replicate only the CPs. The same df (3) would've been generated by replicating three unique runs, chosen at random. From two responses at a given setup, you get one measure, or degree of freedom, to estimate error. In other words, as a general rule: df = n − 1, where n is the subgroup (or

---

sample) size. Replicating CPs, versus any others, at this stage made most sense because it filled the gap in the middle of the factorial space and enabled a test for curvature (described in the section titled "A Test for Curvature..."). It pays to put the power of replication at this location, around which the factors are varied high and low. By replicating the same point a number of times at random, you also create a variability "barometer" (remember those instruments for pressure with the big easy-to-read dials?). Without doing any fancy statistics, you can simply keep an eye on the results of the same setup repeated over and over, similar to the "control" built in as a safeguard against spurious outcomes in classical experimentation.

A Test for Curvature Tailored to 2^k Designs with CPs

We now apply a specific test for curvature that's geared for 2^k designs with CPs. Table 3.3 shows the results of this test, which produces a highly significant p-value of 0.0010.

Let's delve into the details on how to test for curvature. If no curvature exists, the CPs (inner) on average will differ only insignificantly from the average of the factorial points (outer). Therefore, to assess curvature, the appropriate F-test is

$$F = \frac{(\bar{y}_{\text{factorial}} - \bar{y}_{\text{center}})^2}{\hat{\sigma}^2((1/N_{\text{factorial}}) + (1/N_{\text{center}}))}$$

Table 3.3   ANOVA Showing Curvature Test

Source	Sum of Squares (SS)	df	Mean Square (MS)	F-Value	p-Value Prob > F
Model	129.45	4	32.36	17.33	0.0019
Curvature	67.33	1	67.33	36.06	0.0010
Residual	11.20	6	1.87		
Lack of fit	*3.58*	*3*	*1.19*	*0.47*	*0.7242*
Pure error	*7.62*	*3*	*2.54*		
Cor Total	207.99	11			

The denominator expresses the variance associated with the squared difference in means as a function of the number (N) of each point type. In the case of the chemical reaction:

$$F = \frac{(81.175 - 86.2)^2}{1.87\left(\frac{1}{8} + \frac{1}{4}\right)} = \frac{25.25}{0.7} = 36$$

The value of 1.87 comes from the mean square residual in the ANOVA (see Table 3.3) above. It's based on 6 df. The critical value of F for the 1 df of variance in the numerator and 6 df in the denominator is approximately 6, so the value of 36.06 represents highly significant curvature (p = 0.0010). After removing this large source of variance (67.33 $SS_{Curvature}$) by essentially appending a term for curvature to the model, the LOF becomes insignificant (p > 0.7). The conclusion from all this is that the factorial model fits the outer points (insignificant LOF), but obviously not the inner CPs (significant curvature).

Further Discussion on Testing for Lack of Fit (LOF)

Statistical software that does not explicitly recognize CP curvature in 2^k designs produces different results on the ANOVA. Table 3.4 displays the results, which look quite different than before in Table 3.3.

Note the big increase in SS for residuals, specifically those labeled as "Lack of fit." All of the variation previously removed to account for curvature at the CP pours into this "bucket." As a result, the model F-value falls precipitously, causing the associated p-value to increase somewhat above

Table 3.4   ANOVA with Curvature Added to Error

Source	Sum of Squares (SS)	df	Mean Square (MS)	F-Value	p-Value Prob > F
Model	129.45	4	32.36	2.88	0.1052
Residual	78.54	7	11.22		
Lack of fit	70.92	4	17.73	6.98	0.0712
Pure error	7.62	3	2.54		
Cor total	207.99	11			

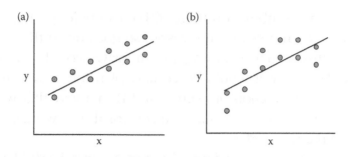

Figure 3.8   (a) LOF not significant. (b) LOF significant.

acceptable levels for significance. Furthermore, since pure error remains unchanged from the prior ANOVA (Table 3.3), the large variance now exhibited for LOF translates to a much more significant p-value of 0.0712. Recall that a similar situation occurred when trying to fit a linear model to the drive-time data: the p-value for LOF fell below 0.1. This is not good.

Figure 3.8a and b illustrates LOF for a simple experiment involving six levels of one factor (x), each of which is replicated for a total of 12 runs.

In Figure 3.8a, an "eye-fit" tells you that all is good. What you see is that the variations of points from the fitted line are about what you'd expect based on variations of the replicate points from their mean (the pure error). This is precisely what's tested for LOF in ANOVA. However, in Figure 3.8b, it's immediately obvious that the line does not fit the points adequately. In this case, the deviations from the fitted line significantly exceed what should be expected from the pure error of replication. Although it's not the only solution, often a better fit can be accomplished by adding more, typically higher order, terms to your model. For example, by adding the x^2 term, thus increasing the model order to quadratic, the LOF would be alleviated in Figure 3.8b, because then you'd capture the curvature. Similarly, in the drive-time example, higher-order terms ultimately got the job done for producing an adequate representation of the true surface.

The Fork in the Road Has Been Reached!

Now we see that a price must be paid for not completing the full 3^3 (=27 run) design for the chemical reaction experiment. ANOVA (in Table 3.3) for the two-level design shows highly significant curvature. However, because all factors, rather than each individually, are run at their mid-levels via CPs, we cannot say whether the observed curvature occurs in the A, B, or C factors or some combination thereof. For example, perhaps the main effect of C is very weak because the response goes up in the middle before dropping off. In this scenario, adding a term for C^2 would capture the evident curvature. However, it would be equally plausible to speculate that the curvature occurs as a function of both A and B in a way that would describe a rising ridge. In this case, the squared terms for these two factors would be added to the predictive model.

Statisticians call this confounding of curvature an "alias" relationship—described mathematically as

$$\text{Curvature} = A^2 + B^2 + C^2$$

Even a statistician cannot separate the squared terms. Table 3.5 illustrates why. It shows the 2^3 design with CPs with factor levels coded according to the standard scheme: minus (–) for low, plus (+) for high, and zero (0) for center. The pattern for the three control factors (A, B, and C) is laid out in standard order. Their effects are calculated using the response data aligned with the pluses versus the minuses.

Higher-order factorial terms such as AB can be estimated by first multiplying the parent terms, in these cases A and B, to create a new column and then calculating the coefficients. However, applying this same procedure to the squared terms does little good, because they all come out the same.

Table 3.5 Aliasing of Squared Terms in Two-Level Factorial with CPs

Std	A	B	C	AB	A^2	B^2	C^2
1	–	–	–	+	+	+	+
2	+	–	–	–	+	+	+
3	–	+	–	–	+	+	+
4	+	+	–	+	+	+	+
5	–	–	+	+	+	+	+
6	+	–	+	–	+	+	+
7	–	+	+	–	+	+	+
8	+	+	+	+	+	+	+
9–12	0	0	0	0	0	0	0

LADIES TASTING TEA VERSUS GIRLS GULPING COLA

The development of modern-day DOE is generally credited to Sir Ronald Fisher (1890–1962), a geneticist who developed statistical tools, most notably ANOVA, for testing crops. He developed the concept of randomization as a protection against lurking variables prevalent in agricultural environments and elsewhere. Fisher discusses randomization and replication at length in his classic article, "Mathematics of a Lady Tasting Tea" (reprinted most recently in Newman's *The World of Mathematics*, Vol. 3, Dover Press, 2000). Fisher's premise is that a lady claims to have the ability to tell which went into her cup first—the tea or the milk. He devised a test whereupon the lady is presented eight cups in random order, four of which are made one way (tea first) and four the other (milk first). Fisher calculates the odds of correct identification as one right way out of 70 possible selections, which falls below the standard 5% probability value generally accepted for statistical significance. However, being a gentleman, Fisher does not reveal whether the lady makes good on her boast of being an expert on tea tasting.

The experiment on tea is fairly simple, but it illustrates important concepts of DOE, particularly the care that must be taken not to confound factors such as what you taste versus run order. For example, as an elementary science project on taste perception, Mark's youngest daughter asked him and several other family members to compare two major brands of cola-flavored carbonated beverages. She knew from discussion in school that consumers often harbor biases, so the taste test made use of unmarked cups for a blind comparison. Unfortunately, this neophyte experimenter put one brand of cola in white Styrofoam® cups and the other in blue plastic glasses in order to keep them straight. Obviously, this created an aliasing of container type with beverage brand. Nothing could be done statistically to eliminate the resulting confusion about consumer preferences. Don't do this at home (or at work)!

> To consult a statistician after an experiment is finished is often merely to ask him to conduct a postmortem examination. He can perhaps say what the experiment died of.
>
> **R.A. Fisher**
> *Presidential Address to the First Indian Statistical Congress, 1938*

It will take more experimentation to pin down the source of curvature in the response from the chemical reaction. In the next chapter we will present part two of this case study, which involves augmenting the existing design with a second block of runs. By adding a number of unique factor combinations, the predictive model can be dealiased in terms of A^2, B^2, and C^2—the potential sources for pure curvature.

PRACTICE PROBLEMS

3.1 From the website from which you downloaded the program associated with this book, open the tutorial titled "*Two-level Factorial*".pdf (* signifies other characters in the file names). You might find several files with this title embedded—pick the one that comes first in sequential order ("Part 1") and follow the instructions. This tutorial demonstrates the use of software for two-level factorial designs. The data you will analyze come from Montgomery (2012). He reported on a waferboard manufacturer who needed to reduce concentration of formaldehyde used as processing aid for a filtration operation. Table 3.6 shows the design sorted by standard order.

You will be well served by mastering two-level factorials such as this before taking on RSM. Obviously, if you will be using Design-Expert® software, the tutorial will be extremely helpful. However, even if you end up using different program, you will still benefit by poring over the details provided on the design and analysis of experiments.

3.2 In Chapter 8 of *DOE Simplified*, we describe an experiment on confetti made by cutting strips of paper to the following dimensions:
   a. Width, inches: 1–3
   b. Height, inches: 3–5

The two-level design with CPs is shown below with the resulting flight times (Table 3.7).

Each strand was dropped 10 times and then the results were averaged. These are considered to be duplicates rather than true replicates because some operations are not repeated, such as the cutting of the paper. On the other hand, we show four separate runs of CPs (standard orders 5–8), which were replicated by repeating all process steps, that is, recutting the confetti to specified dimensions four times, not simply redropping the same piece four times. Thus, the variation within CPs provides an accurate estimate of the "pure error" of the confetti production process. Set up and analyze this data. Do you see significant curvature?

Table 3.6   Data from Montgomery Case Study

Std	A: Temp. (degree Celcius)	B: Pressure (psig)	C: Conc. (percent)	D: Stir Rate (rpm)	Filtration Rate (gallons/hour)
1	24	10	2	15	45
2	35	10	2	15	71
3	24	15	2	15	48
4	35	15	2	15	65
5	24	10	4	15	68
6	35	10	4	15	60
7	24	15	4	15	80
8	35	15	4	15	65
9	24	10	2	30	43
10	35	10	2	30	100
11	24	15	2	30	45
12	35	15	2	30	104
13	24	10	4	30	75
14	35	10	4	30	86
15	24	15	4	30	70
16	35	15	4	30	96

Table 3.7   Flight Times for Confetti Made to Varying Dimensions

Std	A: Width (inches)	B: Length (inches)	Time (seconds)
1	1	3	2.5
2	3	3	1.9
3	1	5	2.8
4	3	5	2.0
5	2	4	2.8
6	2	4	2.7
7	2	4	2.6
8	2	4	2.7

FLYING OFF TO WILD BLUE YONDER VIA "METHOD OF STEEPEST ASCENT"

Box and Liu (1998) report on a fun and informative series of experiments involving optimization of paper helicopters. They apply standard strategies for DOE with a bit of derring-do en route—the method of steepest ascent (Box and Wilson, 1951). This is an aggressive approach that involves taking some risky moves into the unknown, but hopefully in the right direction. It is normally done only at the early phases of process or product improvement—when you are still at the planar stage, far from the peak of performance. If a linear model suffices for predictive purposes, the procedure is made easy with the aid of response surface plots: make moves perpendicular to the contours in the uphill or downhill direction, whichever you seek for optimum outcome. Figure 3.9 illustrates this for an experiment on paper helicopters reported by Box and Liu.

In their example, an initial two-level screening design produces the following linear model of mean flight times (in centiseconds):

$$\hat{y} = 223 + 28x_2 - 13x_3 - 8x_4$$

Only the three factors shown came out significant, out of eight originally tested.

The path of steepest ascent can now be calculated by starting at the center of the original design and changing factors in proportion to the coefficients of the coded fitted equation. In the case of the helicopter, the path was moved along a vector that went up 28 units in x_2 (wing length), down 13 units in x_3 (body length), and down 8 units in x_4 (body width). Box and Liu took off in this direction with redesign of their helicopters and found that flight times peaked at the third step out of five (see Figure 3.10).

They stopped at this point and did a full, two-level factorial on the key factors. Eventually, after a series of iterative experiments, culminating with RSM, a helicopter is designed that flies consistently for over 4 seconds.

The method of steepest ascent is just one element in what Box and Liu call "...a process of iterative learning [that] can ...start from different places and follow different routes and yet have a good chance of converging on some satisfactory solution..."

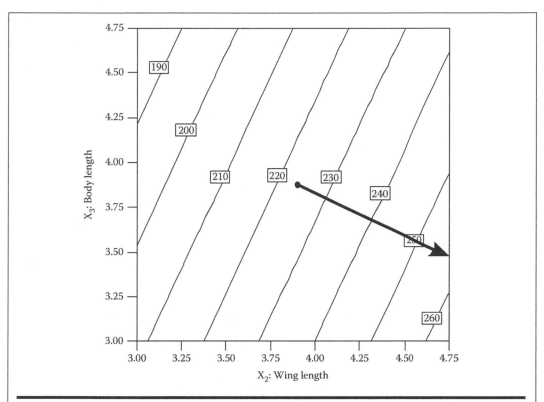

Figure 3.9   Direction of steepest ascent.

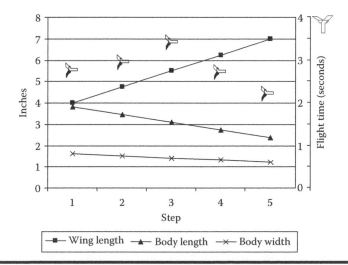

Figure 3.10   Step-by step helicopter dimensions (left axis) versus flight times (right axis).

Chapter 4

Central Composite Design: Stars Added—RSM Show Begins

Two roads diverged in a wood, and I—I took the one less traveled by, And that has made all the difference.

Robert Frost
From The Road Not Taken

The preceding chapters have all been opening acts for the main event—DOE geared for RSM. Assume that you are now approaching the optimum level for your response. Therefore, due to resulting curvature, two-level factorial designs will no longer provide sufficient information to adequately model the true surface. We must dig in at this stage and do more exploration via additional experimental runs at new levels in your factors.

Augmenting a Two-Level Factorial with Star Points to Fit Curvature

We now must deal with a fork in the road created by the discovery of significant curvature in response. The path usually taken in such cases requires running new factor combinations, typically drawn as stars, along the axes of the x space, resulting in a CCD. It's important to keep in mind that this design is intended for sequential experimentation, not a one-shot approach,

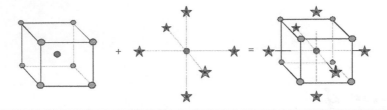

Figure 4.1 Build-up of CCD for three factors.

thus making it flexible for industrial process development. Figure 4.1 shows (for three factors) how the CCD is built up from:

1. Two-level factorial design (plus center points throughout)
2. Axial points represented by stars (plus more center points)

The result, at the far right of Figure 4.1, is a composite design suited to RSM.

The stars in the design pictured in Figure 4.1 go outside of the factorial box along the A (x_1), B (x_2), and C (x_3) axes. This has advantages and disadvantages, which we will discuss when we dissect the CCD in the next chapter. Suffice it to say (for now!) that going further out improves the estimation of curvature. However, if you find it difficult to hit the specified levels, feel free to round them up or down a bit. You can even pull the stars all the way back to the factorial ranges, thus producing a face-centered CCD (called an "FCD") as shown in Figure 4.2.

We put a number of concentric circles around the center points on the FCD to indicate replication and emphasize their role in the construction of

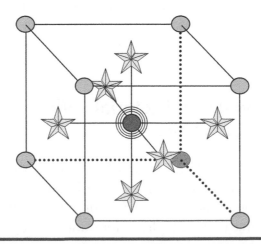

Figure 4.2 Face-centered CCD.

a good RSM design. Obviously, the replication helps for estimation of error, but in sequential experimentation it also provides a link from one block of runs to the next.

BLOCKING OUT KNOWN SOURCES OF VARIATION

Blocking screens out noise caused by known sources of variation, such as raw material batch, shift changes, machine differences, or simply unknown variables that change with time. By dividing your experimental runs into homogeneous blocks, and then arithmetically removing the difference, you increase the sensitivity of your DOE (see e.g., the case study beginning on page 32 of *DOE Simplified* with the section titled "Blocking Out Known Sources of Variation"). A golden rule for good experimentation is: block what you can and randomize what you cannot. You never know what's lurking out there waiting for an opportunity to obscure the true effects, or worse yet, bias your results.

> There are known knowns. These are things we know that we know. There are known unknowns. That is to say, there are things that we know we don't know. But there are also unknown unknowns. These are things we do not know we don't know.
>
> **Secretary of Defense Donald Rumsfeld**
> **on the war against terrorism**

CCD Example: Part Two of Chemical Reaction Example

In the previous chapter, we discussed an experiment aimed at optimizing yields from a chemical reaction. Statistical analysis revealed a significant response increase, in other words: curvature, at the center of the factor (x) space bounded by:

a. Time, minutes: 80–100
b. Temperature, degrees Celsius: 140–150
c. Rate of addition, milliliters/minute: 4–6

Following the procedure for building a CCD illustrated graphically in Figure 4.1, the chemist now completes a second block of runs shown in Table 4.1 (in standard, not random, order). To make it more obvious which combinations are the stars, we designated these rounded values with

Table 4.1 Second Block of Runs on Chemical Reaction

Std Ord	A: Time (minute)	B: Temp. (degree Celsius)	C: Rate (milliliters/ minute)	Yield (grams)
13	73.2* (<−1)	145 (0)	5 (0)	84.3
14	106.8* (>+1)	145 (0)	5 (0)	76.9
15	90 (0)	136.6* (<−1)	5 (0)	77.8
16	90 (0)	153.4* (>+1)	5 (0)	81.6
17	90 (0)	145 (0)	3.32* (<−1)	84.1
18	90 (0)	145 (0)	6.68* (>+1)	86.3
19	90 (0)	145 (0)	5 (0)	88.4
20	90 (0)	145 (0)	5 (0)	87.8

asterisks (*) and noted whether they fall below (<−1) or above (>+1) the factorial box.

How to Select a Good Predictive Model

Recall that the first block, listed in Table 3.2, was composed of eight two-level factorial combinations plus four center points, for a total of 12 runs. Of the second-order terms, it could only fit 2FIs. Squared terms as a whole could be estimated, thus providing an assessment of curvatures, but individual ones remained aliased. Now, with the additional star-point runs, the responses can be modeled by the complete quadratic equation and the source of curvature uncovered:

$$\hat{y} = \beta_0 + \beta_1 A + \beta_2 B + \beta_3 C + \beta_{12} AB + \beta_{13} AC + \beta_{23} BC + \beta_{11} A^2$$
$$+ \beta_{22} B^2 + \beta_{33} C^2$$

However, before generating RSM graphics based on the full quadratic, we'd better go back several steps, starting with the mean model and working up from there. Recall from the note "How Statisticians Keep Things Simple" in Chapter 1 the admonishment that predictive models be kept as parsimonious as possible. You may remember this better as KISS or Keep It Simple, Statistically.

We now introduce a nifty analysis process, called "sequential model sum of squares (SMSS)" (Oehlert, 2000, pp. 56–57), for accomplishing this

Table 4.2   Sequential Model Sum of Squares (SMSS)

Source	Sum of Squares (SS)	df	Mean Square (MS)	F Value	p-Value Prob > F
Mean	$1.380*10^5$	1	$1.380*10^5$		
Block	1.45	1	1.45		
Linear	64.13	3	21.38	1.17	0.3551
2FI	101.19	3	33.73	2.33	0.1258
Quadratic	158.72	3	52.92	32.01	<.0001
Cubic (aliased)	*4.74*	*4*	*1.18*	*0.58*	*0.6889*
Residual	10.14	5	2.03		
Total	$1.384*10^5$	20	6917.64		

KISS principle. Like the ANOVA, the SMSS provides an accounting of variation and associated p-values (Prob > F) so you can see how far it's worth going in degree of polynomial. The objective is to add a higher-level source of terms only if it explains a significant amount of variation beyond what's already accounted for. Table 4.2 reports the results for the entire reaction dataset (block 1 in Table 3.2 and block 2 in Table 4.1).

Appendix 4A provides details on all the calculations in the SMSS, but perhaps it will suffice to simply talk you through it with the focus on interpretation of the outputs. Let's begin with the first SS, labeled "Mean" for the source: this is easily computed by squaring the average of all response values and multiplying this by the number of runs (N). Next, you see a correction for block variation, which in this case did not amount to much. Now we begin with something more substantial in the model—the three linear terms A, B, and C. On the whole, these terms are not sufficient (p of 0.3551 > 0.1). The same conclusion applies to the 2FIs: AB, AC, and BC—not sufficient. However, the quadratic terms (A^2, B^2, and C^2) as a group *are* significant, highly so ($p \ll 0.05$). Therefore, we underlined this model for provisional selection.

NECESSARY, BUT NOT SUFFICIENT

We were careful in saying that the linear and 2FI terms were not sufficient for inclusion in the model. The reason is that the error term at each of these stages still contains variation that can be explained by higher-order terms—in this case the quadratic. (The flowchart provided in Appendix 4A

---

on the SMSS may help you see that this is so.) Therefore, it would be a mistake to say that either class of terms is not significant. As you will see later when we show the ANOVA, all three linear (A, B, and C) terms plus one 2FI term (AB) turn out to be significant at the 0.05 probability level.

Even if all of the linear terms were insignificant according to ANOVA, one or more of them would be included in the final model to maintain model hierarchy. Consider this as family structure imposed on the polynomial. In other words, parents must always be included with their children. For example, if you put the AB interaction in your model, include both A and B individually, even if one or both these linear terms exhibits high p-values (i.e., insignificant). Similarly, A^2 must be supported by A, B^2 by B, and so on. Things get really wacky when a third-order interaction such as ABC is modeled. This term requires not only A, B, and C for hierarchy, but also AB, BC, and AC. In a nutshell, by not maintaining this family structure in your model, some of the statistics may come out wrong under certain circumstances. To be more precise, any polynomial that excludes hierarchically inferior terms will not be invariant to coding (Peixoto, 1990; Nelder, 1998). By maintaining well formulated, hierarchical models you will avoid this and other statistical problems.

We now must test our provisional model for LOF. Notice from Table 4.2 that adding cubic terms would not significantly improve the model (p of $0.6889 \gg 0.1$). Even if it did, the CCD lacks the design points needed to fit all terms required for the cubic, thus it's labeled as being aliased. However, it's nice to know that the chemist evidently did not miss anything by making use of the CCD, which is geared to fit a full quadratic model and nothing more. The LOF tests in Table 4.3 offer more explicit evidence on the adequacy of the quadratic model for approximation of the true surface.

Table 4.3   Lack of Fit Tests

Source	SS	df	MS	F-Value	p-Value Prob > F
Linear	266.98	11	24.27	12.45	0.0132
2FI	165.79	8	20.72	10.63	0.0184
Quadratic	7.07	5	1.41	0.73	0.6395
Cubic (aliased)	2.34	1	2.34	1.20	0.3350
Pure error	7.80	4	1.95		

Refer to Appendix 4B for this chapter for statistical details on how the F-values are computed for LOF. What you should focus on are the associated p-values, which for this test ideally will be high. Since this is a measure of risk, which is a personal matter, you must be the judge, but we advise a healthy dose of skepticism for any models that exhibit LOF below 0.1. In this case, only the quadratic model meets the suggested criterion (p of $0.63055 \gg 0.1$). Don't be confused by how the prior table on model selection was done sequentially: the models listed under "Source" for LOF are whole. Thus, the quadratic now includes the lower-order linear and 2FI terms, so we've got good support for making use of this entire model for predicting the response.

DEPENDENCY OF LOF ON PROPER ESTIMATE OF PURE ERROR

Notice in Table 4.3 that pure error forms the base for LOF tests. In Appendix 4B, you will see that it appears in the denominator of the F-value. Now, recall the sidebar earlier about a chemist who laid out a well-constructed RSM design to be run in random order by a technician, who cheated by dry-labbing the replicated runs. Perhaps, this could be chalked up to a misunderstanding, but it represents an extreme in the continuum of problems in estimates of pure error. More common are the mistakes of simply retesting or resampling rather than rerunning a replicate in the same sequence as every other experimental run. However, the most insidious aspect of getting accurate estimates of pure error is that the person doing the experiments gets very good at doing the same set-up over and over, and worse yet, probably knows what the results were in the past. Then it becomes extremely tempting to abandon an obviously discrepant replicate. Any of these mistakes can lead to an underestimated pure error, which biases the LOF test, making it more significant than it actually should be. Therefore, we advise that you not get overly alarmed at persistent LOF if all other model statistics and diagnostics look good, and most important, when it proves to be useful for predictive purposes.

An error gracefully acknowledged is a victory won.

Caroline Gascoigne

All that remains now is a final assessment of all the models by various measures such as R^2 and PRESS. We talked about these in earlier chapters. Table 4.4 shows the results of this final exam.

Table 4.4   Model Summary Statistics

Source	Std. Dev.	R^2	R^2_{Adj}	R^2_{pred}	PRESS
Linear	4.28	0.1892	0.0271	−0.4636	469.03
2FI	3.80	0.4878	0.2317	−0.4191	480.93
Quadratic	1.29	0.9561	0.9122	0.7873	73.30
Cubic (aliased)	1.42	0.9701	0.8923	−4.5078	1866.64

The quadratic model comes out best for standard deviation, which equals the root mean square (MS) of the residual left over after correcting for the mean, adjusting for blocks and fitting all terms. Obviously, you'd like lower values for this measure. The next three columns in the model summary are different versions of R^2 statistics, all of which should be maximized. The raw R^2 seems to favor the cubic model, but this is a manifestation of the bias discussed previously: adding more model terms invariably inflates this statistic. Don't forget—the added cubic terms were already shown to be insignificant in Table 4.2 on SMSS. Besides, some terms in the cubic model will be aliased, so why waste time evaluating it? Fortuitously in this case, the adjusted R^2 leads you to the simpler quadratic model, which really stands out in the best R^2 of all: predicted. All other models go negative on R^2_{pred}! The final column in Table 4.4 lists the PRESS. This measure, which must be minimized, was already taken into account by R^2_{pred} so it provides no new information, but some analysts may prefer it as a good overall indicator for the best model.

Inspecting the Selected Model

Table 4.5 shows the ANOVA for the model that looks best for predicting reaction yields.

Comparing this to the ANOVA from the first block of runs shown in Table 3.3, you will see that the line labeled "Curvature" is gone, but with the additional block of runs, we can now assess curvature more precisely via the squared terms A^2, B^2, and C^2. The surface exhibits significant ($p < 0.0001$) nonlinearity in both factors A and B, but factor C shows only marginal curvature ($0.05 < 0.0813$ p-value for $C^2 < 0.1$). The 2FI terms are also tested, but only AB creates a significant effect on the response ($p < 0.0001$). The other 2FIs, AC and BC, can safely be eliminated from the model based on their ANOVA statistics, but whether this should be done is debatable (see sidebar "Are You a 'Tosser' or a 'Keeper'?").

Table 4.5 ANOVA for Data from CCD on Chemical Reaction

Source	Sum of Squares (SS)	df	Mean Square (MS)	F-Value	p-Value Prob > F
Block	1.45	1	1.45		
Model	324.04	9	36.00	21.79	<0.0001
A	38.22	1	38.22	23.12	0.0010
B	16.90	1	16.90	10.22	0.0109
C	9.02	1	9.02	5.46	0.0443
AB	99.41	1	99.41	60.15	<0.0001
AC	0.50	1	0.50	0.30	0.5957
BC	1.28	1	1.28	0.77	0.4017
A^2	75.60	1	75.60	45.74	<0.0001
B^2	98.05	1	98.05	59.33	<0.0001
C^2	6.37	1	6.37	3.85	0.0813
Residual	14.87	9	1.87		
Lack of fit	7.07	5	1.41	0.73	0.6395
Pure error	7.80	4	1.95		
Cor Total	340.36	19			

ARE YOU A "TOSSER" OR A "KEEPER"?

Have you observed that, whenever two people operate in close proximity, one will become obsessively neat, tossing anything considered superfluous, and the other keeps any stuff that may conceivably have value? This can be a big source of friction for office- or housemates. Similarly, controversy persists as to the utility of clearing out insignificant model terms not required to maintain hierarchy (family structure). This is easy enough to do with various algorithms for reduction, as noted in Chapter 2 note "A Brief Word on Algorithmic Model Reduction," but is it right or wrong? Here's what some of the experts on DOE say:

In response surface work it is customary to fit the *full* model...

Myers and Montgomery, 2002, p. 742

Choose the smallest order such that no significant terms are excluded.

Oehlert, 2000, p. 56

Our view is to go even further than either of these experts advise, that is, remove any insignificant terms (p > 0.1) not required to maintain model hierarchy. This becomes more and more worthwhile as the number of factors goes up and the order of the model increases. Eliminating insignificant terms then provides big improvements in the R-squared predicted (R^2_{pred}).

Also, we advise you be on the lookout for cases where a factor and every one of its dependents (e.g., A, AB, AC, and A^2) are *all* insignificant. Eliminating these terms would then reduce the dimensionality of the surface, thus making interpretation easier. If this happens to you, watch out, because it begs the question: Why didn't you screen out this nonfactor before doing an in-depth optimization experiment?

All will come out in the washing.

Cervantes (from *Don Quixote*)

The final model, kept whole as suggested by the experts on RSM, is

$$\hat{y} = 86.95 - 1.67A + 1.11B + 0.81C - 3.53AB - 0.25AC + 0.40BC$$
$$- 2.29A^2 - 2.61B^2 - 0.66C^2$$

PREDICTIVE MODELS IN CODED VERSUS ACTUAL UNITS: WHICH IS BEST?

The model shown above works only if you convert factors to the standard coding scale of −1 to +1 for the low versus high end, respectively, of the factorial ranges—defined earlier as: 80–100 minutes of time (A), 140–150 degrees Celsius of temperature (B), and 4–6 milliliters/minute rate of addition (C). With some simple mathematics, the coding can be reversed to translate the model into actual units:

$$Yield = -3205 + 14.3 \, Time + 36.4 \, Temperature - 1.89 \, Rate$$
$$- 0.0705 \, Time * Temperature - 0.025 \, Time * Rate$$
$$+ 0.080 \, Temperature * Rate - 0.0229 \, Time^2$$
$$- 0.104 \, Temperature^2 - 0.665 \, Rate^2$$

We rounded the coefficients in this actual equation to save on space. Depending on the units of measure, round-off error may create serious problems, so be sure to carry all coefficients to many more decimal places than needed for the coded equation. Also, due to their dependency on units, coefficients in the actual equation do not tell you anything.

Coding the factors removes their units of measure, that is, it puts them all on the same scale. The intercept in coded values represents the center of the DOE, and the regression coefficients tell you how the response changes relative to this point of reference. Thus, the coded model facilitates knowledge of your process. However, you may choose to report the uncoded model to make it easier for clients who want to "plug and chug" with actual units of measure.

Regardless of the form of model, coded or actual, do not extrapolate except to guess at conditions for the next set of experiments. The model is only an approximation, not the real truth. It's good enough to help you move in proper direction, but not to make exact predictions, particularly outside the actual experimental region.

P.S. Our models do not include a coefficient for the blocks because it represents variables that cannot be controlled, such as time and materials. However, these effects −0.39 and +0.39 units of yield for blocks 1 and 2, respectively—must be accounted for when diagnosing residuals for abnormalities. This is merely a technicality, but one that cannot be overlooked.

At this stage, it's difficult not to press ahead and generate the response surfaces that will tell the story about how to maximize yields. Don't succumb to this temptation without first diagnosing your residuals for abnormalities. They should be independent of each other and distributed according to a normal distribution with constant variance. Figure 4.3 diagrams how the observed values are filtered for signal by the statistical analysis. The model that results is then used to predict values that, when compared to the original data, produce the residuals.

Examine these residuals to look for patterns that indicate something other than noise is present. If the residuals are pure noise (i.e., they contain no signal), then the analysis is complete.

A quick but effective tool for diagnosing residuals is the normal plot of residuals as shown in Figure 4.4 for our case study.

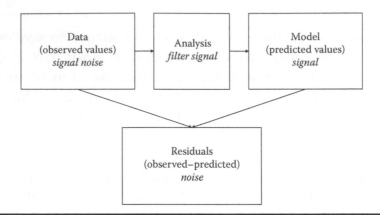

Figure 4.3   Derivation of residuals.

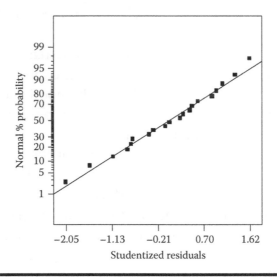

Figure 4.4   Normal plot of residuals for whole model.

This plot lines up nicely as expected from a normal distribution (see *DOE Simplified*, Chapter 1, "Graphical Tests Provide Quick Check for Normality" for details). Here's a simple but effective way to assess the normal plot of residuals: put a pencil over the points. If the pencil covers all the points, everything lines up normally. This is called the "pencil test"!

LEVERAGING YOURSELF INTO TIGHT-FITTING RESIDUALS

Notice that these residuals are studentized. You may recall that in Chapter 2, we made use of externally studentized residuals as a deletion diagnostic for detecting outliers. The residuals plotted above are internally studentized; in other words, none of the actual response data are deleted prior to calculating their deviation from the model prediction. The studentization is essential for accurately diagnosing residuals because it adjusts for varying leverage in design points. Those with high leverage get fitted much tighter than points at the low end of this measure of influence. This must be dealt with in the CCD because the star-point leverage (~0.7) is more than thrice that of the center points (~0.2). Therefore, you can assume from here on out that when we refer to residuals we really mean *internally studentized* residuals but got tired of typing all these words over and over.

P.S. A newer school of thought (Vining, 2011) suggests that <u>all</u> residual diagnostics be externally studentized, the reason being that these follow a t distribution whereas the internally studentized residuals do not. This making little difference in our cases, we left well-enough alone. However, if you use the version of software tied to this book, expect to see residuals externally studentized by default, with an option to revert them to internally studentized or raw (actual units of measure) scale.

Figure 4.5 displays a plot of residuals versus predicted response that provides a handy diagnostic for nonconstant variance.

In this case, the pattern exhibits the hoped-for random scatter. Watch for a megaphone (<) pattern, where the residuals increase with the predicted

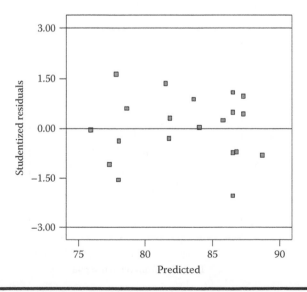

Figure 4.5   Plot of residuals versus predicted response level.

value. In such cases (to be discussed later), consider transforming the responses with a log or possibly the square root.

ANOVA and other statistical analyses are relatively robust to deviations from normality and constancy of variance. Therefore, you should not over-react to slight nonlinearity on the normal plot of residuals, or vague patterns on the residuals versus predicted plot.

There are many more types of diagnostic plots that can and should be produced, including ones discussed earlier that detect outliers and influential combinations of factors, but you may move on with our assurance that all is normal in the reactor case.

Model Graphs

We will now produce contour plots and 3D renderings of the predicted response as a function of the experimental process factors. However, only two factors can be included on any given graph. In cases like the one we're studying that include three or more factors, the perturbation plot helps set priorities on what to graph first. It shows the effects of changing each factor while holding all others constant. Figure 4.6 displays the perturbation plot for the reactor study.

We added a representation of the factor (x) space that shows the paths taken to produce the curves for A, B, and C. A steep slope or curvature in the resulting trace indicates sensitivity of the response to that factor. For

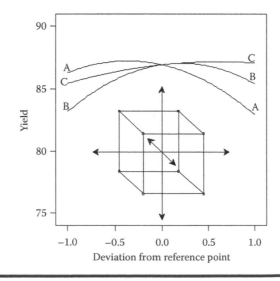

Figure 4.6   Perturbation plot.

example, yield apparently changes more due to A and B than C (the flattest curve).

POSITIVE IS DOWN AND NEGATIVE IS UP

The perturbation plot shows at a glance that all three factors, A, B, and C, curve up and then down. This is caused by the negative coefficients on the squared terms in the model, −2.29, −2.61, and −0.66, respectively. This is counter-intuitive—one would think that these terms must be positive to create the increased response at the center. However, to get this right you must consider that in the coded equation the factors being plugged to the squared terms will be at −1, 0, and +1. Therefore, positive coefficients push up either end, and negative ones press them down, all the while with the middle being unaffected due to the coefficient being zeroed out. So it turns out that, so far as curvature is concerned, positive is down and negative is up.

Note that the paths emanate from the center point. This reference point on the perturbation plot can be changed. The final optimum makes a good reference point from which one can see how sensitive the response becomes to the process factors. The curves may look quite different due to these changes in perspective. For example, see in Figure 4.7 what happens to the perturbation plot after moving factor A down to its −1 factorial level.

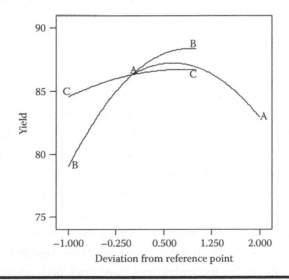

Figure 4.7   Perturbation plot with factor A set low.

Here you see that factors B and C go from −1 to +1 deviation from the adjusted reference point, but factor A now travels two coded units to the right.

We don't recommend you read too much into perturbation plots since they depend so much on the reference point. This plot should be used with caution because it only looks at one-dimensional paths through a multifactor surface. However, it may point out relatively influential factors, in this case: A and B, which make good choices for the axes on the more relevant contour and 3D response surface plots. The other factors must be held constant on any given plot, but to complete the picture you can create several "slices" at various levels, such as setting factor C at −1 to +1 as well as its default value at the center point.

THE INVENTION OF CONTOUR PLOTS

In 1774, a mathematician named Charles Hutton worked on a project to survey Schiehallion mountain in the central Scottish Highlands. The goal was to find its weight for a gravitational experiment aimed at determining the mass of Earth. The surveyors recorded scores of elevations on their map of the mountain. Hutton noticed that by connecting similar values with penciled lines, the slopes and overall shape of the surface became much clearer. Thus, he invented contour lines.

Bill Bryson
A Short History of Nearly Everything,
New York: Broadway Books, 2003, p. 57.

Figure 4.8 displays a contour plot of the yield as a function of A (time) and B (temperature). Each curve represents a region of constant response. However, you mustn't believe what you see outside the box defined by the factorial ranges, in this case 80–100 for A and 140–150 for B.

It would be tempting to predict the response all the way out to the axial extremes represented by the star points. This is a common mistake made by experimenters who use the CCD as a template. The problem is that the precision of the prediction degrades very quickly beyond the −1 to +1 range. This is illustrated graphically via the standard error (SE) plots in Figure 4.9a and b.

The circular patterns of SE are no accident: they result from the careful construction of the CCD by its architects Box and Wilson, who desired equal precision equally distant from the center point (details to be provided later).

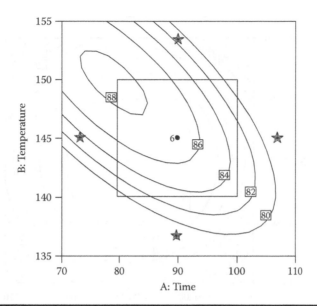

Figure 4.8   Two-dimensional (2D) plot A versus B (C at midpoint).

Statisticians deem designs like this "rotatable" because they can be rotated around their center point without changing the prediction variance (PV) at any given distance in the x-space.

More important, the contour plot of SE (Figure 4.9a) exhibits a broad open-space around the center point (labeled 6 to indicate the replication: 4 in the first block and 2 in the second). This region translates to the flat bottom of the bowl on the 3D rendering (Figure 4.9b). The error increases steeply beyond the factorial range, so predictions beyond that will be relatively poor. Therefore, we advise that you stay inside the box defined by the −1 to +1 factorial levels when producing graphs of the responses. (When searching for the optimum you can be a bit bolder as noted below.)

GOING OUTSIDE THE BOX

For those who like to think outside the box, that is, beyond the plus or minus 1 coded unit limits, a more aggressive region for prediction follows the SE contour that intersects the factorial points, for example, the SE 1.0 contour in Figure 4.9a. For a standard CCD, this defines a circular or spherical region as you can see in the figure.

Taking such a bold, but possibly very productive, approach is facilitated by software that supports numerical optimization within specified

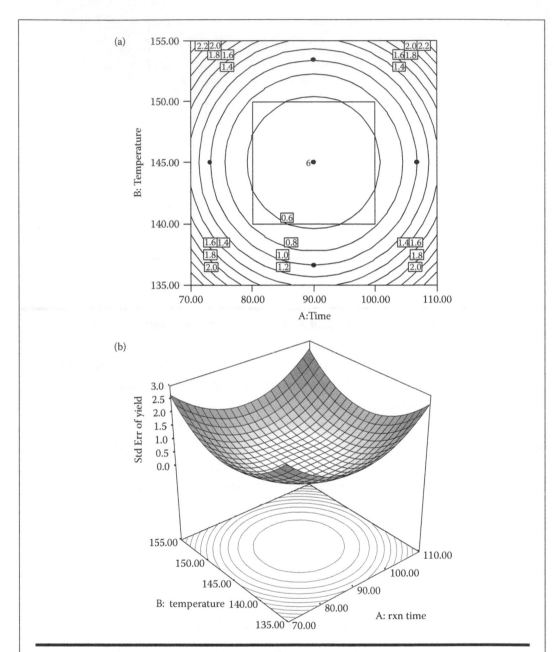

Figure 4.9   (a) Contour plot of SE and (b) 3D plot of SE.

SE ranges. Then you simply set an upper limit at the maximum value seen on the SE contour plot at the factorial extremes. Given this venturesome exploration present no variability beyond what would be experienced at the corners of the experiment, it really is not that risky.

Space: the final frontier. These are the voyages of the starship *Enterprise*. Its continuing mission: to explore strange new worlds, to seek out new life and new civilizations, to boldly go where no one has gone before.

***Star Trek: The Next Generation*, narrated by Patrick Stewart**

Heeding our own advice, we generated the 3D response plots in Figure 4.10a and b with the factors held within their factorial range. These graphs show A versus B with factor C fixed at its −1 and +1 coded levels, which correspond to actual values of 4–6 milliliters per minute rate of addition. Practitioners of RSM would say that they plotted A and B with slices on C. (This brings to mind a big block of Swiss cheese being cleaved evenly along one dimension to reveal interesting patterns on the interior.)

We have not displayed the plot for C set at its center point, but as you've seen in ANOVA, the model and the perturbation plot, this factor has little impact on response, so it would be hard to distinguish from the two we already show above. By inspection of graphs (and verified by numerical optimization), the peak yield can be found at:

a. Time: 80 minutes (−1 coded level)
b. Temperature: 150 degrees Celsius (+1)
c. Rate of addition: 6 milliliters/minute (+1)

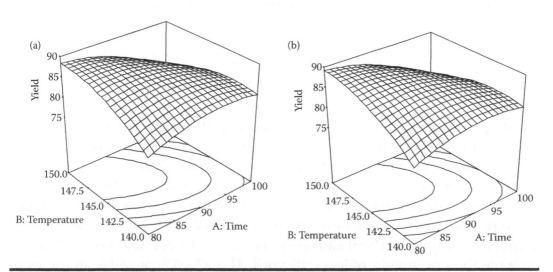

Figure 4.10 (a) 3D plot with C at −1 and (b) 3D plot with C at +1.

ALGORITHMS FOR OPTIMIZATION

The term "algorithm" is an eponym for Al-Khwarizm, a ninth century Persian mathematician. While working as a scholar at the House of Wisdom in Baghdad, he wrote a book entitled *Al-jabr wa'l muqabala* from which our modern word "algebra" comes. This book introduced the decimal system as well as rules for solving equations, including the quadratic. Our interest in the quadratic equations stems from its ability to model response surfaces. Later in the book, we will introduce a hill-climbing algorithm that can be easily programmed for use on a computer. In conjunction with a good objective function for multiple responses, it handles a variety of goals desired by industrial experimenters.

We live in the best of all possible worlds.

Gottfried Leibniz
Seventeenth century inventor of calculus
and the concept of optimization

To generate the predicted response at these factor levels, simply plug them into the quadratic model:

$$\hat{y} = 86.95 - 1.67A + 1.11B + 0.81C - 3.53AB - 0.25AC + 0.40BC - 2.29A^2$$
$$- 2.61B^2 - 0.66C^2$$
$$= 86.95 - 1.67(-1) + 1.11(+1) + 0.81(+1) - 3.53(-1)(+1)$$
$$- 0.25(-1)(+1) + 0.4(+1)(+1) - 2.29(-1)^2 - 2.61(+1)^2 - 0.66(+1)^2$$
$$= 86.95 + 1.67 + 1.11 + 0.81 + 3.53 + 0.25 + 0.4 - 2.29 - 2.61 - 0.66$$
$$= 89.15$$

The predicted value of 89.1 agrees well with the actual outcome at the same setup of factors: 88.2 from standard order 7 (shown in Table 3.2). However, it always pays to do a follow-up run for confirmation. Obviously, some variation must be expected due to normal variation in the process and the response measurements. Further allowances must be made due to inherent imprecision that comes from the limited sampling done for experimental purposes. Fortunately, the data used to fit the model can also be used to compute the uncertainty of its predictions. The ANOVA provides an estimate of variance via the MS of the residuals. Via the appropriate statistical formula (see sidebar titled "Formulas for Standard Error [SE] of Predictions"), a prediction interval (PI) can be constructed to convey uncertainties about confirmation runs done after the actual experiment. In this case, the PI for 95% confidence is 85.4–92.9 around the expected outcome of 89.1 grams of yield. In other words, any individual outcome within this interval does not invalidate the model. The width of the PI may raise a few eyebrows, but keep in mind that the starting point is merely an educated guess based on one set of runs—then come the normal variations of a lined-out, that is, stable, process. Even though people may be unpleasantly surprised by the PI, we advise that you report it in order to manage expectations. Then you're less likely to get shot down when results come out a bit different than expected.

FORMULAS FOR SE OF PREDICTIONS

Good regression software will take care of this for you, but for those who want all the details on calculations, here are formulas for CIs on predictions from simple linear regression (one factor x):

$$\hat{y} = \beta_0 + \beta_1 x_*$$

$$SE_{mean(*)} = \hat{\sigma}\sqrt{\frac{1}{n} + \frac{(x_0 - \bar{x})^2}{\sum_{i=1}^{n}(x_i - \bar{x})^2}}$$

$$SE_{pred} = \hat{\sigma}\sqrt{1 + \frac{1}{n} + \frac{(x_0 - \bar{x})^2}{\sum_{i=1}^{n}(x_i - \bar{x})^2}} = \sqrt{\hat{\sigma}^2 + (SE_{mean(*)})^2}$$

where the SE is an abbreviation for standard error and

- sigma-hat (σ with ^) is the estimated standard deviation that you can estimate via the square root of the residual MS from ANOVA
- n is the number of runs
- x_* represents the point being predicted.

(These formulas, with more details and explanation than we offer here, can be found in academic textbooks on applied linear regression such as Weisberg, Section 1.7.)

The 95% PI is then calculated in the usual way by making use of multipliers from Student's t-table (*DOE Simplified*, Appendix 1.1) for the df listed on ANOVA for the residual. For example, referring back to Table 4.5 for the full model, we find df of 9, which produces a two-tailed t-value (the multiplier) of 2.262. The SE of prediction (calculated via regression software) is 1.66 at the optimum reaction conditions (−1A, +1B, and +1C) that produced a response (Y) of 89.1 grams yield. Thus, the PI becomes 89.1± (2.26) 1.66, which agrees closely (some round-off error occurs in our calculations) with the range of 85.4–92.9 generated by software and reported in the text.

By following the road map for CCD shown in Figure 4.1, which adds stars to a core factorial design, we made all the difference in modeling curvature. See this for yourself by comparison of the before (factorial-only) and after (full CCD) response surfaces in Figure 4.11 with center points depicted.

Notice how the quadratic surface made possible by completing the CCD (Figure 4.11b) cuts through the middle of the center points, whereas with only a 2FI (2FI model, shown in Figure 4.11a) the actual results all differ from predicted.

Like the road taken by Robert Frost, the path from factorial design to CCD for RSM is less traveled, but very rewarding for experimenters.

PRACTICE PROBLEMS

4.1 From the website for the program associated with this book, open the software tutorial entitled "*Multifactor RSM*".pdf (* signifies other

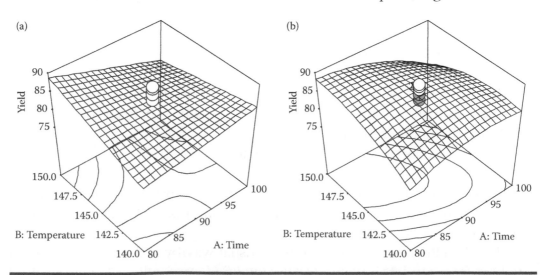

Figure 4.11   (a) Factorial response surface and (b) CCD response surface.

characters in the file name). You might find several files with this title embedded—pick the one that comes first in sequential order (it should say "Part 1") and follow the instructions. This tutorial demonstrates the use of software for setting up and analyzing a CCD done on a chemical process. Table 4.6 shows the experiment sorted in standard order.

This tutorial exercises a number of features in the software, so if you plan to make use this tool for RSM, it should not be bypassed. However, even if you end up using different software, you will still benefit by poring over the details of the design and analysis of experiments.

Table 4.6   CCD on a Chemical Process

Std	Block	A: Time (minute)	B: Temp. (degree Celsius)	C: Catalyst (%)	Conversion (%)	Activity
1	Day 1	40	80	2	74	53.2
2	Day 1	50	80	2	51	62.9
3	Day 1	40	90	2	88	53.4
4	Day 1	50	90	2	70	62.6
5	Day 1	40	80	3	71	57.3
6	Day 1	50	80	3	90	67.9
7	Day 1	40	90	3	66	59.8
8	Day 1	50	90	3	97	67.8
9	Day 1	45	85	2.5	81	59.2
10	Day 1	45	85	2.5	75	60.4
11	Day 1	45	85	2.5	76	59.1
12	Day 1	45	85	2.5	83	60.6
13	Day 2	36.6	85	2.5	76	53.6
14	Day 2	53.4	85	2.5	79	65.9
15	Day 2	45	76.6	2.5	85	60.0
16	Day 2	45	93.4	2.5	97	60.7
17	Day 2	45	85	1.66	55	57.4
18	Day 2	45	85	3.34	81	63.2
19	Day 2	45	85	2.5	80	60.8
20	Day 2	45	85	2.5	91	58.9

Table 4.7   Second Block of Runs to Optimize Flight Time of Confetti

Std	A: Width (inches)	B: Length (inches)	Time (seconds)
9	0.6	4.0	2.5
10	3.4	4.0	1.8
11	2.0	2.6	2.6
12	2.0	5.4	3.0
13	2.0	4.0	2.5
14	2.0	4.0	2.6
15	2.0	4.0	2.6
16	2.0	4.0	2.9

4.2 This is a continuation of the experiment on confetti started in the previous chapter—see Table 3.7 (originally reported in Chapter 8 of *DOE Simplified*). ANOVA on the results from the two-level factorial design with center points reveals significant curvature. Therefore, the experimenters add a second block of runs to create a CCD. The result can be seen in Table 4.7.

Points 9 through 12 in standard (Std) order represent the stars placed approximately 1.4 coded units from the center along the two axes A and B. As you can decipher from the factor levels, the remaining four runs are center points that help to link this second block to the first (shown in Table 3.7). Analyze the entire set of data for confetti and determine the source of the curvature in response. Then find the optimal configuration of length and width (within the experimental bounds) to maximize flight time.

Appendix 4A: Details for Sequential Model Sum of Squares Table

The SMSS in Table 4.2 provides a comparison of models. It shows the statistical significance of adding new model terms step-by-step in increasing order:

$$\hat{y} =$$

$$1 \quad \beta_0$$

$$2 \quad + \beta_{blk_1} + \beta_{blk_2}$$

$$3 \quad + \beta_1 x_1 + \beta_2 x_2 + \beta_3 x_3$$

$$4 \quad + \beta_{12} x_1 x_2 + \beta_{13} x_1 x_3 + \beta_{23} x_2 x_3$$

$$5 \quad + \beta_{11} x_1^2 + \beta_{22} x_2^2 + \beta_{33} x_3^2$$

$$6 \quad + \beta_{111} x_1^3 + \cdots + \beta_{112} x_1^2 x_2 + \cdots + \beta_{123} x_1 x_2 x_3 \quad (10 \text{ terms})$$

The table begins with the total sum of squares (SS). This is the variation of the responses using zero as the baseline. The first polynomial term, the intercept β_0, accounts for the mean. Its SS is the first to be set aside in the sequence shown above. Next comes the adjustment for blocks, if the experimenter included these in the design. At this stage as each succeeding layer of terms comes into the model it is tested (via F) against the remainder of the SS.

Figure 4A.1 illustrates how the sequential F-values were computed for Table 4.2 according to the procedure outlined here.

These calculations help you select the highest-order model where the terms are significant and not aliased.

Appendix 4B: Details for Lack of Fit Tests

Again for reference sake, we've diagrammed in Figure 4B.1 how the LOF F-values are computed.

The first few layers in this diagram make adjustments to the total sum of squares (SS) for the mean and blocks. After that, models are assessed for LOF in increasing order—linear, 2FI, quadratic, and cubic. The pure error SS, which comes from replicated runs, is separated from the raw residuals after each model. The remainder is designated as LOF. Its MS is divided by the MS of the pure error to generate the F-value for LOF. A significant F-value can indicate the numerator still contains signal. This leads to exploration of higher-order terms that model any remaining signal and thus remove it from the residuals.

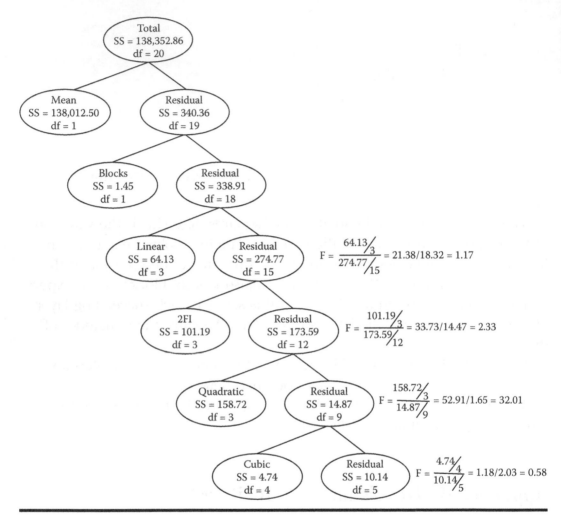

Figure 4A.1 Calculations for SMSS in Table 4.2.

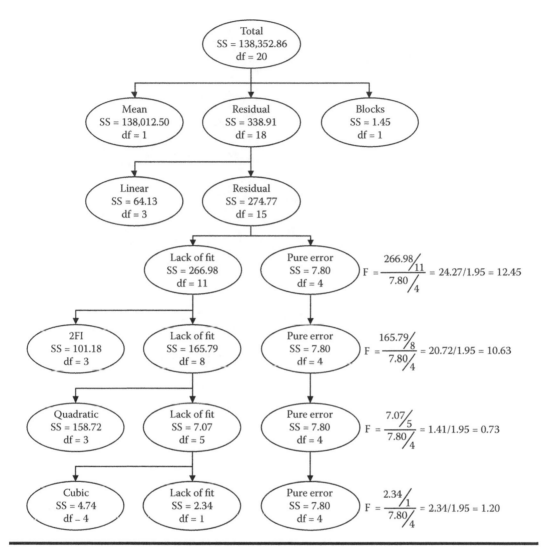

Figure 4B.1   Calculations for LOF results shown in Table 4.3.

Chapter 5

Three-Level Designs

Two is company, three is trumpery.

Jane G. Austin

Betty Alden: The First-born Daughter of the Pilgrims, 1891, p. 171.

In the previous chapter, we introduced the CCD—the first choice for practitioners of RSM because of its flexibility. As illustrated by a case study, an experimenter can hedge on RSM by adding CPs to a simpler, more economical two-level factorial. If curvature is not significant, it's time to move on: why bother doing RSM if the surface is planar? On the other hand, if you detect a significant increase or decrease in response at the CP *and* it merits further attention, you can complete the CCD by adding the second block of axial points, including more CPs to provide a link with block one.

MUCH ADO ABOUT NOTHING?

In the previous chapters, we detailed a study on a reaction process that revealed a significant curvature effect of about 6 grams above the expected level of 82 grams. The chemist evidently felt that this deviation was too important to ignore, so the second block of the CCD was completed in order to construct a proper map of the nonlinear response. If the curvature had been overlooked despite being statistically significant, it would've made no difference in the end, because the same optimum emerges from the factorial model as that produced by the full CCD: low time, high temperature, and high rate. Thus, we get into an issue of statistical significance versus practical importance. Highly controlled automated assay optimization equipment now can accommodate DOEs with hundreds of reactions

---

run at varying times, temperatures, plus other process factors and changing chemical compositions (Erbach et al., 2004). These massive designs have the statistical power to reveal tiny effects—some so small that they are of no practical importance. If you see the word *significant* trumpeted in technical reports, don't be misled into thinking the research must then be considered important. Look at the actual effects generated by the experiment in relation to what should be considered of economic or other value.

Let every eye negotiate for itself and trust no agent; for beauty is a witch against whose charms faith melteth in blood.

Shakespeare
Much Ado about Nothing (II, i, 178–180)

The axial (star) points ideally (according to the developers Box and Wilson) go outside of the factorial box. This has advantages and disadvantages. It's good to go further out for assessing curvature. However, it may be inconvenient for the experimenter to hit the five levels required of each factor: low axial (star at smallest value), low factorial, CP, high factorial, and high axial (star at greatest value). Furthermore, the stars may break the envelope of what's safe or even physically possible. For example, what if you include a factor such as a dimension or a percent and the lower star point comes out negative? In such cases, the experimenter may opt to do a face-centered CCD (FCD) as shown in Figure 4.2.

Now let's suppose you are confident that a two-level factorial design will not get the job done because you're already in the peak region for one or more of the key process responses. The obvious step is upgrading to a three-level factorial (3^k). However, as discussed in Chapter 3, though this would be a good choice if you've narrowed the field to just two key factors, beyond that the 3^k becomes wasteful. Fortunately for us, George Box put his mind to this and with assistance from Box and Behnken (1960) came up with a more efficient three-level design option.

FEELING LUCKY? IF SO, CONSIDER A "DEFINITIVE SCREENING DESIGN"

In the third edition of *DOE Simplified*, we introduce definitive screening designs (DSDs) as near-minimal run (2K + 1) resolution IV templates for uncovering main effects, their novelty being a layout with three (not just two) levels of each factor (Jones and Nachtsheim, 2011). Because DSDs generate squared terms, they serve as a response surface method. However, you must have at hand the right computational tools for deriving a model from a design with fewer runs than the number of coefficients in the model, that is, a "supersaturated" experiment.

This not being a robust choice for an RSM design, we recommend you take the DSD short cut only under extenuating circumstances. If you do, apply forward selection (mentioned in Chapter 2's side note "A Brief Word on Algorithmic Model Reduction") using the Akaike (pronounced ah-kah-ee-keh) information criterion (AICc), rather than the usual p values. The small "c" in the acronym refers to a correction that comes into play with small samples to prevent overfitting.

An easygoing overview on AICc is provided by Snipes and Taylor in a case study (2014) applying this criterion to the question of whether the more you pay for wine the better it gets. For reasons which will become apparent, consider pouring yourself a chilled Chardonnay before downloading this freely available (under a Creative Commons license) publication from www.sciencedirect.com/science/article/pii/S2212977414000064.

The Box–Behnken Design (BBD)

BBDs are constructed by first combining two-level factorial designs with incomplete block designs (IBD) and then adding a specified number of replicated CPs.

INCOMPLETE BLOCK DESIGNS

An IBD contains more treatments than can be tested in any given block. This problem emerged very early in the development of DOE for agricultural purposes. For example, perhaps only two out of three varieties of a crop could be planted at low and high levels in any of three possible fields. In such a case, any given field literally represented a block of land within which only an incomplete number of crop varieties could be tested. Incomplete blocks may occur in nonagricultural applications as well. Let's say you want to compare eight brands of spark plugs on a series of six-cylindered engines in an automotive test facility. Obviously, one engine block can only accommodate an incomplete number of plugs (six out of the eight).

**Table 5.1   Component Elements of
Three-Factor BBD**

a	b
−1	−1
+1	−1
−1	+1
+1	+1

A	B	C
a	b	0
a	0	b
0	a	b
0	0	0

Table 5.1 provides the elements for building a three-factor BBD. On the left, you see a two-level factorial for two factors: a and b. Wherever these letters appear in the incomplete block structure at the right, you plug in the associated columns to create a matrix made up of plus 1s (high), minus 1s (low), and 0s (center).

So, for example, the first run in standard order is −1, −1, 0. Then comes +1, −1, 0, and so forth, with +1, +1, 0 being the last of the first group of four runs. The following four runs will all be set up with B at the 0 level. After four more runs with A at zero the design ends with a CP—all factors being at the zero level. At least three CP runs are recommended for the BBD (Myers et al., 2016, p. 413). However, we advise starting with five and going up from there for designs with more factors. Table 5.2 shows the end result: 17 runs including five replicated CPs.

Figure 5.1 shows the geometry for this three-factor BBD, which displays 12 edge points lying on a sphere about the center (in this case at $\sqrt{2}$) with five replicates of the CP.

The 13 unique combinations represent less than one-half of all possible combinations for three factors at three levels ($3^3 = 3 * 3 * 3 = 27$) and yet they provide sufficient information to fit the 10 coefficients of the quadratic polynomial.

LESSON ON GEOMETRY FOR BBD

Although the description is accurate for three factors, it is misleading to refer to edge points as a geometric element. That's why our picture lays out the points as corners of squares (consistent with the incomplete block structure). The BBD uses only −1, 0, and +1 for the factor levels, so the design points are actually the centroids of the (k − m) dimensional faces of the k-dimensional cube, where m is the number of ±1's in a design row and k represents the number of factors. For the special case of the three-factor BBD, k is 3 and m is 2, so the design points fall on the center of the one (k − m = 3 − 2 = 1) dimensional faces—that is, the centers of the cubical edges.

BBDs are geared to fit second-order response surfaces for three or more factors. The BBDs are rotatable (for k = 4, 7) or nearly so. Table 5.3 shows how many runs, including CPs, are needed as a function of the number of factors up to seven.

As shown above, most BBDs can be safely subdivided into smaller blocks of runs. Templates for these designs and for many more factors are available

Table 5.2 Three-Factor BBD

Std	A	B	C
1	−1	−1	0
2	+1	−1	0
3	−1	+1	0
4	+1	+1	0
5	−1	0	−1
6	+1	0	−1
7	−1	0	+1
8	+1	0	+1
9	0	−1	−1
10	0	+1	−1
11	0	−1	+1
12	0	+1	+1
13	0	0	0
14	0	0	0
15	0	0	0
16	0	0	0
17	0	0	0

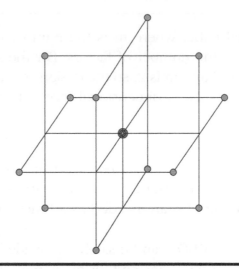

Figure 5.1   Layout of points in a three-factor BBD.

Table 5.3   Sampling of BBDs

Factors k	BBD Runs (CPs)	BBD Blocks
3	17 (5)	1
4	29 (5)	1 or 3
5	46 (6)	1 or 2
6	54 (6)	1 or 2
7	62 (6)	1 or 2

via the Internet (Block and Mee) and encoded in RSM software (Design-Expert, Stat-Ease, Inc.). Bigger and bigger BBDs, accommodating 20 or more factors, are currently being developed by the statistical academia (Mee, 2003).

Case Study on Trebuchet

The following case study makes use of a scale-model trebuchet built by the South Dakota School of Mines and Technology (SDSMT) for experimentation by their engineering students. Figure 5.2 shows it being made ready to fire a rubber racquetball.

The ball is held in a sling attached to a wooden arm via fishing line. When the arm is released by the trebuchet operator, the counterweights lever it upward—causing the ball to be flung forward 100 feet or more.

Figure 5.2   SDSMT trebuchet.

Notice the holes drilled through the wooden arm, which allow various configurations for placement of the weights and the pivot point. Here's what we studied:

A. Arm length: 4–8 inches (in.) from counterweight end to pin that held weights
B. Counterweight: 10–20 pounds (lbs)
C. Missile weight: 2–3 ounce (oz.) racquetball filled with varying amounts of salt

---

QUICK PRIMER ON MEDIEVAL
MISSILE-HURLING MACHINERY

Catapults, energized by tension or torsion, may be more familiar to you than the trebuchet (pronounced "treb-you-shay"), which uses a counterweight. The first trebuchets were built prior to the fifth century B.C. in China (Chevedden, 1995). The technology migrated to Europe where machines like this used for war became known as engines (from the Latin *ingenium* or ingenious contrivance). Operators of trebuchets were called "ingeniators"—a precursor to a profession now known as engineering.

Britain's King Edward, villainously depicted in the movie *Braveheart,*
kept a whole crew of ingeniators busy bombarding the Scottish-held
Stirling Castle in 1304 with 13 trebuchets going day and night. They
saved their biggest engine of war, called the "War Wolf," for the coup de
grâce—firing several missiles despite efforts by the Scots to surrender
first. Trebuchets used a variety of ammunition, mainly stone balls, but
also beehives, barrels of pitch or oil that could be set ablaze, animal car-
casses to spread diseases, and hapless spies—captured and repatriated
as the crow flies.

A heavy silence descended … I stepped back to contemplate "the
beast." She was truly magnificent—powerful, balanced, of noble
breed.

Renaud Beffeyte
An ingeniator for trebuchets reproduced at Loch Ness, Scotland, for
a public television show by NOVA (Hadingham, 2000)

These three key factors were selected based on screening studies performed by SDSMT students (Burris et al., 2002). The experiments were
performed by one of the authors (Mark) and his son (Hank) in the backyard
of their home. To make the exercise more realistic, they aimed the missiles
at an elevated play fort about 75 feet from the trebuchet. It took only a few
pre-experimental runs to determine what setting would get the shots in
range of the target. At first, the ingeniators (see sidebar titled "Quick Primer
on Medieval Missile-Hurling Machinery" for explanation of term) tried
heavier juggling balls (1–2.5 pounds), but, much to their chagrin, these flew
backward out of the sling. The lighter racquetballs worked much better—flying well beyond 75 feet over the fence and into the street. (Beware to neigh-
bors strolling with their babies by the Anderson household!)

Based on the prior experiments and physics (see the "Physics of the
Trebuchet" note), it made no sense to simply do a two-level factorial design:
RSM would be needed to properly model the response. It would have been
very inconvenient to perform a standard CCD requiring five-factor levels.
Instead, the experimenters chose the BBD specified in Table 5.2. The ranges
for arm length (factor A) and counterweight (B) were chosen to be amenable
with the SDSMT trebuchet. These factors were treated as if they could be
adjusted to any numerical level, even though in practice this might require
rebuilding a trebuchet to the specified design.

PHYSICS OF THE TREBUCHET

The trebuchet is similar to a first-order lever, such as see-saws or teeter-totters you see at the local playgrounds. Force is applied to one end, the load is on the other end, and the fulcrum sits between the two. Imagine putting a mouse on one end of the see-saw and dropping an elephant on the other end—this illustrates the physics underlying the trebuchet. These principles were refined for centuries by ancient war-mongers. For example, in the fourteenth century, Marinus Sanudus advised a 1:6 ratio for the length of the throwing arm to the length of the counterweight arm, both measured from the point of the fulcrum (Hansen, 1992). Arabic sources suggested that to get maximum effect from basketfuls of poisonous snakes, the length of the sling should be proportional to the length of the throwing arm. Physics experts can agree on the impact of these variations in length because their effects on leverage are well known. However, they seem to be somewhat uncertain about other aspects of trebuchet design, such as

■ Adding wheels to the base (evidently helps by allowing weight to drop more vertically as the trebuchet rolls forward in reaction to release)

■ Hinging the weight rather than fixing it to arm (apparently this adds an extra kick to the missile)

The results from the 17-run BBD on the SDSMT trebuchet are shown in Table 5.4. The response (y) was distance measured in feet.

Standard order 9 (run number 14) actually hit the elevated fort, so the short distance it would have traveled beyond that was estimated at 8 feet (so total = 75 + 8 = 83). Notice from the last five runs in standard order, all done at CP levels, how precisely the trebuchet threw the racquetball. Because it was tricky to get an exact spotting for landings, the measurements were rounded to the nearest foot, so the three results at 91 actually varied by a number of inches, but that isn't much at such a distance. The relatively small process variation, in comparison to the changes induced by the controlled factors, led to highly significant effects. You can see this in the sequential model sums of squares statistics in Table 5.5.

Table 5.4   Results from Trebuchet Experiment (Underlined Shot Hit the Target)

Std	Run	A	B	C	y
1	13	4	10	2.5	33
2	12	8	10	2.5	85
3	4	4	20	2.5	86
4	7	8	20	2.5	113
5	2	4	15	2.0	75
6	17	8	15	2.0	104
7	9	4	15	3.0	40
8	8	8	15	3.0	89
9	14	6	10	2.0	83
10	3	6	20	2.0	108
11	11	6	10	3.0	49
12	6	6	20	3.0	101
13	1	6	15	2.5	88
14	5	6	15	2.5	91
15	15	6	15	2.5	91
16	10	6	15	2.5	87
17	16	6	15	2.5	91

Table 5.5   Sequential Model Sums of Squares

Source	Sum of Squares (SS)	df	Mean Square (MS)	F-Value	p-Value Prob > F
Mean	$1.176 * 10^5$	1	$1.176 * 10^5$		
Linear	7236.75	3	2412.25	34.70	<0.0001
2FI	438.50	3	146.17	3.14	0.0738
Quadratic	437.7	3	145.92	37.21	0.0001
Cubic (aliased)	12.25	3	4.08	1.07	0.4543
Residual	12.25	4	3.80		
Total	$1.258 * 10^5$	17	7397.18		

HOLLYWOOD TAKES ON THE TREBUCHET

To get a look at a trebuchet in action, rent or stream the video for *The Last Castle* (2001) starring Robert Redford (the hero, naturally) and James Gandolfini as a sadistic warden. Be forewarned that this movie, like many produced in Hollywood, features a number of highly improbable occurrences, which could only happen in an alternate universe in which the laws of physics, probability, and common sense no longer apply. For example, Redford supervises construction of a collapsible trebuchet under the noses of the guards, using only the materials available in the prison workshop and then trains his crew to use it during exercise breaks, so when the climactic riot occurs, they can then fling a rock through the warden's window for a direct hit on to his prized display case of weaponry. According to one reviewer (Andrew Howe, *FilmWritten*, November 20, 2001. http://www.efilmcritic.com/review.php?movie=5566&reviewer=193), *The Last Castle* is "a jaw-dropping, bone-crunching mess of a film, the product of one too many mind-altering substances." However, it's very inspiring if you aspire to be a master ingeniator with a trebuchet.

Table 5.5 shows that linear terms significantly improve the fit over simply taking the mean of all responses. The 2FI terms appear to fall into the gray area between p of 0.05 and 0.10, but be careful. These probability values are biased due to an error term that contains variability, which will be explained by the next order of terms. We talked about this in a note titled "Necessary, But Not Sufficient" in the previous chapter. Be sure to look again at these 2FI terms at the ANOVA stage of the analysis. There's no doubt about the quadratic terms—they are highly significant (p of 0.0001). The BBD does not support estimation of all cubic terms so this model is aliased and shouldn't be applied.

The quadratic model appears to adequately represent the actual response surface based on its insignificant lack of fit (p >> 0.1) shown in Table 5.6.

Table 5.6   LOF Tests

Source	SS	df	MS	F	p-Values Prob > F
Linear	888.52	9	98.72	25.98	0.0034
2FI	450.02	6	75.00	19.74	0.0061
Quadratic	12.25	3	4.08	1.07	0.4543
Pure error	15.20	4	3.80		

Overall, the quadratic model emerges best according to the model summary statistics listed in Table 5.7.

Table 5.8 displays the ANOVA for the full quadratic model.

All quadratic terms except for B^2 achieve significance at the 0.05 p-value threshold (95% CI). We will carry this one insignificant term along with the rest. The residuals from this full quadratic model exhibit no aberrant patterns on the diagnostic plots discussed in previous chapters (normal plot of residuals, residuals versus predicted level, etc.), so we now report the results.

Table 5.7   Model Summary Statistics

Source	Std. Dev.	R^2	R^2_{Adj}	R^2_{Pred}	PRESS
Linear	8.34	0.8890	0.8634	0.7939	1678.11
2FI	6.82	0.9429	0.9086	0.8155	1502.20
Quadratic	1.98	0.9966	0.9923	0.9730	219.75
Cubic (aliased)	1.95	0.9981	0.9925		

Table 5.8   ANOVA

Source	Sum of Squares (SS)	df	Mean Square (MS)	F-Value	p-Value Prob > F
Model	8113.02	9	901.45	229.88	<0.0001
A	3081.13	1	3081.13	785.71	<0.0001
B	3120.50	1	3120.50	795.76	<0.0001
C	1035.13	1	1035.13	263.97	<0.0001
AB	156.25	1	156.25	39.85	0.0004
AC	100.00	1	100.00	25.50	0.0015
BC	182.25	1	182.25	46.48	0.0002
A^2	364.17	1	364.17	92.87	<0.0001
B^2	4.64	1	4.64	1.18	0.3126
C^2	45.85	1	45.85	11.69	0.0111
Residual	27.45	7	3.92		
Lack of fit	12.25	3	4.08	1.07	0.4543
Pure error	15.20	4	3.80		
Cor Total	8140.47	16			

Before generating the more interesting 2D contour and 3D response surfaces, it will be helpful to take a simplistic view of how the three control factors affected the distance off the trebuchet. This can be seen on the perturbation plot shown in Figure 5.3.

As you can infer from B^2 not being significant according to ANOVA, the track for factor B exhibits no curvature (nonlinearity). On the other hand, factor A exhibits a noticeable bend. Factor C exhibits only a very slight curve. Only two of these three factors can be included on contour and 3D plots. Obviously, A should be one of them because it's the most arresting visually. Figure 5.4a and b shows the contour plot and corresponding

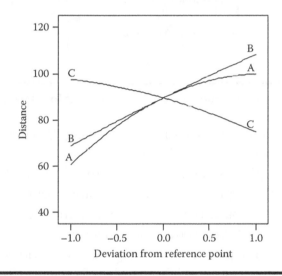

Figure 5.3   Perturbation plot.

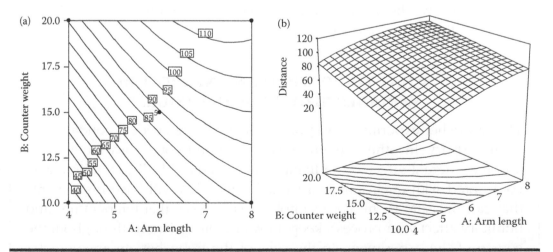

Figure 5.4   (a) 2D plots (C centered). (b) 3D plot (C centered).

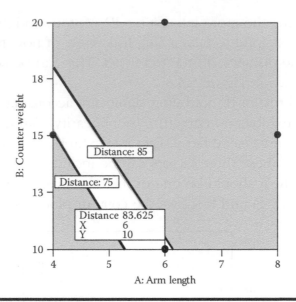

Figure 5.5 Graphical overlay plot (factor C constant at low level).

3D view for A (arm length) versus B (counterweight) while holding C (missile weight) at its CP value (2.5 ounces).

Notice on the 2D plot in Figure 5.4a how the contours for 75–85 feet pass diagonally roughly through the middle. This range represents the operating window or sweet spot. It puts the 2.5-ounce missile somewhere on the target at a broad array of A–B setup combinations.

The graphical overlay plot shown in Figure 5.5 provides a compelling view of the sweet spot at a different slice for factor C—its lower level (2.0-ounce missile).

The flag shows the setup used at standard order 9 listed on Table 5.4, which actually hit the fort as predicted.

WHAT TO DO WHEN AN UNEXPECTED VARIABLE BLOWS YOU OFF COURSE

The trebuchet experiment was done outdoors under ideal conditions—calm and dry. In other circumstances, the wind could have built up to a point where it affected the flight of the missile. Ideally, experimenters confronted with environmental variables like this will have instruments that can measure them. For example, if you know that temperature and humidity affect your process, keep a log of their variations throughout the course of the experiment. In the case of the trebuchet, an anemometer would be handy for logging wind magnitude and direction. If you think such a variable should be accounted for, add it to the input matrix as a covariate. Then, its contribution to variance can be accounted for in the ANOVA, thus revealing controlled factors that might otherwise be obscured.

Hornblower told himself that a variation of two hundred yards in the fall of shot from a six pounder at full elevation was only to be expected and he knew it to be true, but that was cold comfort to him. The powder varied from charge to charge, the shots were never truly round, quite apart from the variations in atmospheric conditions and in the temperature of the gun. He set his teeth, aimed and fired again—short and a trifle to the left. It was maddening!

C.S. Forester
Flying Colours (1938)

PRACTICE PROBLEM

5.1 RSM experiments on gasoline engines help automotive companies predict fuel consumption, exhaust emissions, etc. as functions of engine rpm, load, and so on. The prediction equations then become the basis for managing engine parameters via on-board computers in cars. For example, Ford Motor Limited of the United Kingdom applied a BBD to maximize fuel (or as they call it, "petrol") efficiency and minimize various exhaust emissions as a function of five key factors (Draper et al., 1994):

A. Engine load, Newton meters (N m): 30–70

B. Engine speed, revolutions per minute (rpm): 1000–4000

C. Spark advance, degrees (deg): 10–30

D. Air-to-fuel ratio: 13–16.4

E. Exhaust gas recycle, percent of combustion mixture (%): 0–10

The automotive engineers divided this 46-run experiment into two blocks, presumably on two test engines. The results for one of the emissions, carbon monoxide (CO), are shown in Table 5.9. Obviously, this should be minimized.

Using the software associated with this book, you can set up this five-factor BBD and enter the results for the CO emissions. (Or save time by opening the file named "Prob 5-1 emissions.dx*" available via the Internet: see Appendix 5A on software installation for a path to

Table 5.9   Engine Experiment

Std	Blk	A: Load (Nm)	B: Speed (rpm)	C: Spark Adv (deg)	D: Air/ Fuel Ratio	E: Recycle (%)	CO (ppm)
1	1	30	1000	20	14.7	5	81
2	1	70	1000	20	14.7	5	148
3	1	30	4000	20	14.7	5	348
4	1	70	4000	20	14.7	5	530
5	1	50	2500	10	13	5	1906
6	1	50	2500	30	13	5	1717
7	1	50	2500	10	16.4	5	91
8	1	50	2500	30	16.4	5	42
9	1	50	1000	20	14.7	0	86
10	1	50	4000	20	14.7	0	435
11	1	50	1000	20	14.7	10	93
12	1	50	4000	20	14.7	10	474
13	1	30	2500	10	14.7	5	224
14	1	70	2500	10	14.7	5	346
15	1	30	2500	30	14.7	5	147
16	1	70	2500	30	14.7	5	287
17	1	50	2500	20	13	0	1743
18	1	50	2500	20	16.4	0	46
19	1	50	2500	20	13	10	1767
20	1	50	2500	20	16.4	10	73
21	1	50	2500	20	14.7	5	195
22	1	50	2500	20	14.7	5	233
23	1	50	2500	20	14.7	5	236
24	2	50	1000	10	14.7	5	100
25	2	50	4000	10	14.7	5	559
26	2	50	1000	30	14.7	5	118

(Continued)

Table 5.9 (*Continued*) **Engine Experiment**

Std	Blk	A: Load (Nm)	B: Speed (rpm)	C: Spark Adv (deg)	D: Air/ Fuel Ratio	E: Recycle (%)	CO (ppm)
27	2	50	4000	30	14.7	5	406
28	2	30	2500	20	13	5	1255
29	2	70	2500	20	13	5	2513
30	2	30	2500	20	16.4	5	53
31	2	70	2500	20	16.4	5	54
32	2	50	2500	10	14.7	0	270
33	2	50	2500	30	14.7	0	277
34	2	50	2500	10	14.7	10	303
35	2	50	2500	30	14.7	10	213
36	2	30	2500	20	14.7	0	171
37	2	70	2500	20	14.7	0	344
38	2	30	2500	20	14.7	10	180
39	2	70	2500	20	14.7	10	280
40	2	50	1000	20	13	5	548
41	2	50	4000	20	13	5	3046
42	2	50	1000	20	16.4	5	13
43	2	50	4000	20	16.4	5	123
44	2	50	2500	20	14.7	5	228
45	2	50	2500	20	14.7	5	201
46	2	50	2500	20	14.7	5	238

the website offering files associated with the book.) After creating the design (or opening the file), we suggest you put it in standard order to enter the data (or look it over)—it will come up by default in random run order. Then, analyze the results.

Watch out for significant lack of fit and problems in the diagnostic plots for residuals, especially the Box–Cox plot (see Appendix 5A). Then, reanalyze the data with a log transformation (previously applied to the American football example in Chapter 2—and discussed in Appendix 2A). If you get bogged down on the transformations or any other aspect of this case, don't belabor the problem: see what we've done in the solution (stored in Adobe's PDF) posted at the same website where you found the data for this problem).

AUTOMOTIVE TERMINOLOGY IN THE UNITED KINGDOM VERSUS UNITED STATES

One of the authors (Mark) had a very pleasurable time driving a rented Ford (United Kingdom) Mondeo while traveling in Ireland. Upon returning to the United States, he learned that Ford sold the same car under the name "Contour," which for an enthusiast of RSM could not be resisted—it was promptly purchased! More than just the brand names change when automotive technology hops over "The Pond" (Atlantic Ocean)—a whole set of terms must be translated from English to American English. For example, the original authors (UK-based) of the case described in Problem 5.1 refer to "petrol" rather than "gasoline," as we in the United States call it. The UK affiliate for Stat-Ease once told a story about how he tracked down a problem in his vehicle, which required looking under the "bonnet" (hood) and checking the "boot" (trunk). His UK-made engine was made from "aluminium" (aluminum). It starts with power from the "accumulator" (battery), which is replenished by the "dynamo" (generator). An indicator for the battery and other engine aspects can be seen on the "fascia" (dashboard). Our man in the United Kingdom checks the air pressure in his "tyres" (tires) with a "gauge" (gage) and holds parts to be repaired in a "mole wrench" (vice grips) or, if it's small enough, a "crocodile clip" (alligator clip). Repairs are done with the aid of a "spanner" (wrench), perhaps by the light of a "torch" (flashlight). When he heads down the "motorway" (highway) dodging "lorries" (trucks), the British driver's engine sounds are dampened by the "silencer" (muffler).

We have everything in common with America nowadays, except, of course, the language.

Oscar Wilde
From his romantic comedy The Canterville Ghost, 1887

Giving English to an American is like giving sex to a child. He knows it's important but he doesn't know what to do with it.

Adam Cooper (nineteenth century)

Appendix 5A: Box–Cox Plot for Transformations

This is as good a time as any to introduce a very useful plot for response transformations called the "Box–Cox" (Box and Cox, 1964). You should make use of this plot when doing Problem 5.1. Figure 5.6 shows a Box–Cox plot for the regression analysis of quarterback (American football) sacks reported in Chapter 2. It was produced by applying a linear (first-order) model: $\hat{y}$ = f (height, weight, years, games, and position).

The plot indicates the current power (symbolized mathematically by the Greek letter lambda) by the dotted line at 1 on the x-axis. This represents no transformation of the actual response data (y^1). Alternatively, the response is transformed by a range of powers from −3 (inverse cubed) to +3 (cubed). The transformed data are then refitted with the proposed model (in this case linear) and the residual SS generated in a dimensionless scale for comparative purposes. (Box and Cox recommended plotting against the natural logarithm (ln) of the residual SS (versus the raw $SS_{residual}$), but this is not of critical importance.) The minimum model residual can then be found (see the tall line in plot) and the CI calculated via statistical formulas (displayed on either side of the minimum with short lines). In this case (football data), notice that the current point (the dotted line) falls outside of the 95% CI. Therefore, applying a different power, one within the CI at or near the minimum, will be advantageous. It's convenient in this case to select a power of 0, which represents the logarithmic transformation, either natural or base 10—it does not matter: the difference amounts to only a constant factor (Box and Draper, p. 289).

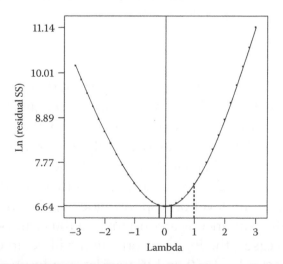

Figure 5.6   Box–Cox plot for football data.

The indication for log transformation (lambda (λ) zero) given by the Box–Cox plot shown in Figure 5.6 is very typical of data that varies over such a broad range (18-fold in this case). However, this is just one member from a family of transformations—designated as power law by statisticians—you should consider for extreme response data. Other transformations that might be revealed by the Box–Cox plot are

- Square root (0.5 power), which works well for counts, such as the number of blemishes per unit area
- Inverse (–1 power), which often provides a better fit for rate data

George Box and his colleagues offer these general comments on transformations, in particular the inverse (Box et al., 1978, p. 240): "The possibility of transformation should always be kept in mind. Often there is nothing in particular to recommend the original metric in which the measurements happen to be taken. A research worker studying athletics may measure the time t in seconds that a subject takes to run 1000 meters, but he could equally well have considered 1000/t, which is the athlete's speed in meters per second."

THE POWER-LAW FAMILY

This might be a good description for a family of high-priced lawyers of several generations, such as "Smythe, Smythe, and Smythe, Attorneys-at-Law." However, it's something even more scary—a statistical description of how the true standard deviation (σ) might change as a function of the true mean (μ). Mathematically, this is expressed as

$$\sigma_y \propto \mu^\alpha$$

Ideally, there is no relationship (the exponent α equals 0), so no transformation is needed (lambda (λ) minus alpha (α) equals 1, so the power-law specifies y^1, which leaves the responses in their original metric). When standard deviation increases directly as a function of the mean ($\alpha = 1$), you may see a characteristic megaphone pattern (looks like: <) of residuals versus predicted response. In other words, the residuals increase directly as the level of response goes up. Then, it will be incorrect to assume that variance is a constant and thus can be pooled for the ANOVA. In such cases, the log transformation will be an effective remedy (because $\lambda = 1 - \alpha = 1 - 1 = 0$ and y^0 translates to $\ln(y)$).

Other transformations, not part of the power-law family, may be better for certain types of data, such as applying the arc-sin square root to fraction-defect (proportion pass/fail) data from quality control records (see Anderson and Whitcomb, 2003, for details).

To summarize this and previous discussion on diagnosing needs for response transformation:

1. Examine the following residual plots (all studentized) to diagnose non-normality (always do this!):
   a. Normal plot
   b. Residuals versus predicted (check on assumption of constant variance)
   c. Externally studentized (outlier t) versus run number
   d. Box–Cox plot (to look for power transformations)
2. Consider a response transformation as a remedy, such as
   a. The logarithm (base 10 or natural, it does not matter)
   b. Another one from the power-law family such as square root (for counts) or inverse (for rates)
   c. Arc-sin square root (for fraction defects) and other functions not from the power-law family

Chapter 6

Finding Your Sweet Spot for Multiple Responses

> We make trade-offs in every solution, trying to minimize the pain
> and maximize efficiency, but never entirely ridding ourselves of the
> pain and never quite building the perfect system.
>
> **Michael F. Maddox**
> *Computer programmer*

Up until now, so as not to complicate matters, we've concentrated only on
one response per experiment. However, any research done for practical purposes invariably focuses on multiple responses. We will now provide tools
to deal with the inevitable tradeoffs that must be made when dealing with
multiple responses from designed experiments.

"Less Filling" versus "Tastes Great!": Always a Tradeoff

In the mid-1970s, Miller Brewing Company introduced its Lite beer brand
with the advertising tagline "less filling/tastes great." Whether they actually
accomplished this difficult tradeoff may be a matter of debate, but it struck
a chord with American beer drinkers. The Miller ads promoted the view
that with proper technology, a sweet spot can be found where all customer
needs are met.

HISTORY OF LIGHT BEER

By the late 1990s, light beers accounted for over one-third of all beers brewed in the United States. Miller, Budweiser, and Coors led the production in this market segment, which amounted to nearly 80 million barrels of beer. Dr. Joseph L. Owades, founder of The Center for Brewing Studies in Sonoma, California, came up with the idea for a low-calorie beer when he worked at Rheingold—a New York brewer. He investigated why some adults did not drink beer (many people don't!). The answers he got were twofold

1. "I don't like the way beer tastes"
2. "I'm afraid it will make me fat"

Owades couldn't do anything about the taste of beer, but he did something about the calories by inventing a process that got rid of all the starch in beer, thus lowering the calories. The enzymes in barley malt only break down to about two-thirds of the starch to the point where yeast can digest it; so, one-third is left behind in the beer as a body. Owades came up with the idea that if you introduced a new enzyme, like people have in their stomachs, it would break down all the starch, convert it into alcohol, and leave a low-carbohydrate, high-alcohol beer. He didn't just take a standard beer and dilute it because that would not taste the same.

Almost 40 years later, brewers do it a little differently. They simply use less malt, thus introducing less-fermentable sugars. The result is beer that has fewer calories, but also less body and malt flavor.

Does modern light beer taste as great as regular beer? You must be the judge.

Priscilla Estes
New Beverage, 1998 editorial at www.beveragebusiness.com

Making Use of the Desirability Function to Evaluate Multiple Responses

How does one accomplish a tradeoff such as beer that's less filling while still tasting great? More likely in your case, the problem is how to develop a process that with maximum efficiency at minimum cost, achieves targeted quality attributes for the resulting product. In any case, to make optimization easy, the trick is to combine all the goals into one objective function. Economists do this via a theoretical utility function that produces consumer satisfaction. Reducing everything to monetary units makes an ideal objective function, but it can be difficult to do. We propose that you use desirability (Derringer, 1994) as the overall measure of success when you optimize multiple responses.

To determine the best combination of responses, we use an objective function, $D(x)$, which involves the use of a multiplicative rather than an additive mean. Statisticians call this the geometric mean. For desirability, the equation is

$$D = (d_1 \times d_2 \times \cdots \times d_n)^{1/n} = \left(\prod_{i=1}^{n} d_i \right)^{1/n}$$

The d_i, which ranges from 0 to 1 (least to most desirable, respectively), represents the desirability of each individual (i) response, and n is the number of responses being optimized. You will find this to be a common-sense approach that can be easily explained to your colleagues: convert all criteria into one scale of desirability (small d's) so that they can be combined into one, easily optimized, value—the big D. This overall desirability D can be plotted with contours or in 3D using the same tools as those used for the RSM analysis for single responses. We will show these when we get to a more detailed case study later in this chapter.

By condensing all responses into one overall desirability, none of them need to be favored more than any other. Averaging multiplicatively causes an outcome of zero if any one response fails to achieve at least the tiniest bit of desirability. It's all or nothing.

WHY NOT USE THE ADDITIVE MEAN FOR INDIVIDUAL DESIRABILITY?

When this question comes up in Stat-Ease workshops, we talk about desirability on the part of our students to study in a comfortable environment. It is often the case that the first thing in the morning of a day-long

class, the temperatures are too cold, but with all the computers running and hot air emitted by the instructors (us!); by afternoon, things get
uncomfortably hot. At this stage, as a joke, we express satisfaction that
on average (the traditional additive mean), the students enjoyed a desirable room temperature (they never think that this is funny ☹). Similarly,
your customers will not be happy if even one of their specifications does
not get met. Thus, the geometric mean is the most appropriate for overall
desirability (D).

Π. Σ. (P. S.) In the equation for D, notice the symbol Π (Greek letter pi, capitalized—equivalent to "p" in English) used as a mathematical
operator of multiplication, versus the symbol Σ (capital sigma—or "s" in
English) used for addition. To keep these symbols straight, think of "s"
(Σ) for sum and "p" (Π) for product.

Now, with the aid of numerical optimization tools (to be discussed in
Appendix 6A), you can search for the greatest overall desirability (D) not
only for responses but also for your factors. Factors normally will be left free
to vary within their experimental range (or in the case of factorial-based
designs, their plus/minus 1 coded levels), but for example, if time is an
input, you may want to keep it to a minimum.

If all goes well, perhaps, you will achieve a D of 1, indicating that you
satisfied all the goals. However, such an outcome may indicate that you're
being a slacker—that is, someone who does not demand much in the way
of performance. At the other extreme, by being too demanding, you may
get an overall desirability of zero. In this case, try slacking off a bit on
the requirements or dropping out some of them altogether until you get
a nonzero outcome. As a general rule, we advise that you start by impos-
ing goals only on the most critical responses. Then, add the less-important
responses, and finally, begin setting objectives for the factors. By sneaking
up to the optimum in this manner, you may learn about the limiting
responses and factors—those that create the bottlenecks on getting a
desirable outcome.

The crucial phase of numerical optimization is the assignment of various
parameters that define the application of individual desirabilities (d_i's). The
two most important are

- Goal: none, maximum, minimum, target, or range
- Limits: lower and upper

Here's where subject matter expertise and knowledge of customer requirements becomes essential. Of lesser importance are the parameters:

- Weight: 1–10 scale.
- Importance: 5-point scale displayed + (1 plus) to +++++ (5 plus).

For now, leave these parameters at the default levels of 1 and +++ (3 plus), respectively. We will get to them later. It is best to start by keeping things as simple as possible. Then, after seeing what happens, impose further parameters such as the weight and/or importance on specific responses (or factors—do not forget that these can also be manipulated in the optimization). First, let's work on defining desirability goals and limits.

If your goal is to *maximize* the response, the desirability must be assigned as follows:

$d_i = 0$ if y (or x) falls below the lower limit
$0 \leq d_i \leq 1$ as y (or x) varies from low to high
$d_i = 1$ if y (or x) exceeds the higher-threshold limit

We say "limit," but these are really thresholds because variables may fall outside these values—in this case, preferably above the higher limit. Figure 6.1 shows how desirability ramps up as you get closer to your goal at the upper threshold.

Hearkening back to the beer example, this is the proper goal for the "tastes great!" boast. Let's say you establish a taste rating of 1 (low) to 10 (high) and ask drinkers what would be satisfactory. Obviously, they'd raise their mugs only to salute a rating of 10. In reality, you would almost certainly get somewhat lower ratings from an experiment aimed at better taste *and* less-filling beer. But in situations like this, if your clients desire the absolute maximum, set the high limit above what you actually observe to put a "stretch" in the objective. For example, an experiment on light beer might produce only a rating of 6 at the best (let's be honest—taste does

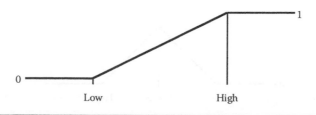

Figure 6.1   Goal is to maximize.

Figure 6.2   Goal is to minimize.

suffer in the process of cutting calories). In this case, we suggest putting in a high limit (or stretch target) of 10 nevertheless, because possibly, the actual result of 6 happened to be at the low end of the normal variability and therefore, more can be accomplished than you may think. Don't rack your brain over this—just do it!

The desirability ramp shown in Figure 6.2 shows the opposite goal—*minimize.*

This is the objective for the "less filling" side of light beer. Notice that the ramp for the minimum is just flipped over from that shown for the maximum. We won't elaborate on the calculations for the d_i because you can figure it out easily enough from the details we provided for the maximum goal. If you want the absolute minimum, set the low limit very low—below what you actually observed in your experimental results; for example, a zero-calorie beer.

Quite often, the customer desires that you hit a specified target. To be reasonable, they must allow you some slack on either side of the goal as shown in Figure 6.3.

Notice how the desirability peaks at the goal. It is a conglomeration of the maximum and minimum ramps. We won't elaborate on the calculations for the d_i—you get the picture. Getting back to the beer, consider what would be wanted in terms of the color. Presumably via feedback from focus groups in the test marketing of the new light variety, the product manager decided on a targeted level of measurable amber tint, which was not too clear (watered-down look) and not too thick (a heavy caloric appearance).

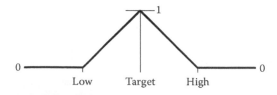

Figure 6.3   Goal is target (specification).

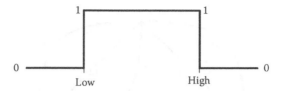

Figure 6.4   Goal is a range.

We mentioned earlier that you can establish goals for factors, but you may want to put this off until a second pass on the overall desirability function. However, unless you ran a rotatable CCD and went for Chapter 4's "Reaching for the Stars" limits set to the standard error of the axial points, we advise you to constrain your factors within the ranges that were experimented upon. Do not extrapolate into poorly explored regions! Figure 6.4 illustrates such a range constraint in terms of individual desirability.

We advise that when running a standard CCD, you set the low and high limits at the minus/plus 1 (coded) factorial levels, even though their extreme values go outside this range. As discussed earlier, the idea behind CCDs is to stay inside the box when making predictions and seeking optimum values. Technical note: According to Derringer and Suich (1980), the goal of a range is simply considered to be a constraint; so, the d_i, although they get included in the product of the desirability function D, are not counted in determining n for $D = (\Pi d_i)^{1/n}$.

Now, we've got the basics of desirability covered; so, let's put the icing on the cake by considering additional parameters—importance and weights.

If you want to finc-tunc your optimization, consider making use of a desirability parameter called "weight," which affects the shapes of the ramps so that they more closely approximate what your customers want. The weight can vary from 0.1 to 10 for any individual desirability. Here's how it affects the ramp

■ =1, the d_i go up or down on straight (linear) ramps this is the default
■ >1 (maximum 10), the d_i get pulled down into a concave curve that increases close to the goal
■ <1 (minimum 0.1), the curve goes to the opposite direction (convex) going up immediately after the threshold limit, well before reaching the goal

Think literally in terms of applying actual weights to flexible ramps of desirability. This is illustrated in Figure 6.5 showing the impact of varying weight on desirability over a targeted specification.

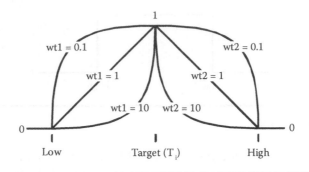

Figure 6.5   Goal is target (specification)—weights added.

Notice how the ramps are pulled down by the highest weights (10) and lifted up by the lower ones (weight of 0.1). Mathematically, this is expressed as

$$d_i = 0 \text{ at } y_i \leq \text{Low}_i$$

$$d_i = \left[\frac{y_i - \text{Low}_i}{T_i - \text{Low}_i} \right]^{wt1_i} \quad \text{when Low}_i < y_i < T_i$$

$$d_i = \left[\frac{y_i - \text{High}_i}{T_i - \text{High}_i} \right]^{wt2_i} \quad \text{when } T_i < y_i < \text{High}_i$$

$$d_i = 0 \text{ at } y_i \geq \text{High}_i$$

where T refers to the targeted value. Similar calculations can be done to apply weights to the goals of the minimum or maximum. Why you'd apply weights is a tougher question. This may be a stretch, but let's say that consumers associate the color of light beer with its lightness in terms of calories. Perhaps then, if the color varied off the target to the higher side, it would cause a big drop in desirability, but on the lower end, it would take a big reduction for drinkers to be put off. In such a case, you'd want to put a high weight at the right (upper end) and a low weight at the left, thus creating the curve shown in Figure 6.6.

Now, we focus our attention on one more parameter for desirability—the importance rating. In the objective function $D(x)$, each response can be assigned a relative importance (r_i) that varies from 1 plus (+) at the least to 5 plus (+++++) at the most. If varying degrees of importance are assigned to the different responses, the objective function is

$$D = \left(d_1^{r_1} \times d_2^{r_2} \times \cdots \times d_n^{r_n} \right)^{1/\Sigma r_i} = \left(\prod_{i=1}^{n} d_i^{r_i} \right)^{1/\Sigma r_i}$$

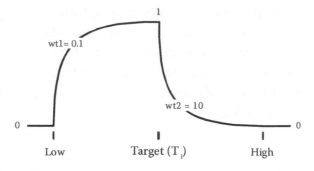

Figure 6.6   Goal is target (specification)—weights added in a peculiar manner.

where n is the number of variables (responses and/or factors) in D(x). When all the importance values remain the same, the simultaneous objective function reduces to the general equation shown earlier (without the r_i exponents needed to account for importance). Change the relative importance ratings only if you believe that some individual desirabilities deserve more priority than others. For example, it seems likely that in the case of light beer, taste would be considered more critical to sales than the reduction in calories, with everything else being equal. Arguably, one could assign importance ratings as follows:

- ■ + (1 plus) for calories
- ■ +++ (3 plus) for color
- ■ +++++ (5 plus) for taste

However, you'd better evaluate the impact of varying importance on the suggested optimum.

Numerical Search for the Optimum

Once you've condensed all your responses down to one univariate function via the desirability approach, optimization becomes relatively straightforward with any number of numerical methods. One unbiased and relatively robust approach involves a hill-climbing algorithm called "variable-sized simplex" (see Appendix 6A). Because more than one hill can exist in desirability space, the optimum found (the hill climbed) depends on where you initiate the search. To increase the chances of finding the global optimum, a number of optimization cycles, at least 100, should be done using different starting points chosen at random or on a grid. It's safe to assume that you will find software with this tool for numerical optimization (e.g., the package accompanying this book) or an alternative approach. Let's not belabor this point, but leave it to experts in computational methods to continue improving the accuracy and precision for consistently finding the global optimum in the least amount of time. This sounds like a multiple-response optimization problem, don't you think?

Summary

The method we recommend for multiple-response optimization follows these steps:

1. Develop predictive models for each response of interest via statistically designed RSM experiments on the key factors.
2. Establish goals (minimize, maximize, target, etc.) and limits (based on customer specification) for each response, and perhaps also factors, to accurately determine their impact on individual desirability (d_i). Consider applying weights to shape the curves more precisely to suit the needs of your customers and applying varying importance ratings, if some variables must be favored over others.
3. Combine all responses into one overall desirability function (D) via a geometric mean of the d_i computed in step 2. Search for the local maxima on D and rank them as best to worst. Identify the factor settings for the chosen optimum.
4. Confirm the results by setting up your process according to the recommended setup. Do you get a reasonably good agreement with expected response values? (Use the prediction interval as a guide for assessing the success or failure to make the confirmation.)

If you measure many responses, the likelihood increases that your mission to get all of them within specification becomes impossible. Therefore, you might try optimizing only the vital few responses. If this proves to be successful, establish goals for the trivial many, but do this one by one so that you can determine which response closes the window of desirability—thus producing no solution.

Some of your responses, ideally not one of those that are most important, may be measured so poorly (e.g., a subjective rating on the appearance of a product) that nothing significant can be found in relation to the experimental factors. The best you can do for such responses will then be to take the average, or mean model (cruel!). Also, do not include any responses whose "adequate precision," a measure of signal to noise defined in the Glossary, falls below 4, even if some model terms are significant.

Case Study with Multiple Responses: Making Microwave Popcorn

In *DOE Simplified* (Chapter 3), we introduced an experiment on microwave popcorn as a primer for two-level factorial design. Two responses were measured; so, this simple case is useful to illustrate how to establish individual desirability functions and then perform the overall optimization.

The popcorn DOE involved three factors, but one—the categorical brand of popcorn did not create a significant effect on any of the responses. That allows us to now concentrate on the other two factors: time and power— both numerical. We will fit factorial models to the multiple-response data shown in Table 6.1 and perform a multiple-response optimization.

To simplify tabulation of the response data, we show two results per row by a standard order for the original 2^3 design reported in *DOE Simplified*, but assume that these were run separately in random order. We made a few other changes in terminology from the previous report:

■ Time and power have been relabeled as "A" and "B"
■ The term "UPK" is brought in to better describe the second response, which is the weight of <u>u</u>n<u>p</u>opped <u>k</u>ernels of corn (previously labeled as "bullets")

Table 6.1 Data for Microwave Popcorn

Std Order	A: Time (minutes)	B: Power (percent)	y_1: Taste (rating)	y_2: UPK (ounces)
1, 2	4 (−)	75 (−)	74, 75	3.1, 3.5
3, 4	6 (+)	75 (−)	71, 80	1.6, 1.2
5, 6	4 (−)	100 (+)	81, 77	0.7, 0.7
7, 8	6 (+)	100 (+)	42, 32	0.5, 0.3

PRIMER ON MICROWAVE OVENS

Microwaves are high-frequency electromagnetic waves generated by a magnetron in consumer ovens. Manufacturers typically offer several power levels—more levels at the top of their product lines, which run upward from 1000 watts in total. The amount of power absorbed by popcorn and other foods depends on its dielectric and thermophysical properties. Water absorbs microwaves very efficiently so that moisture content is critical. Microwave energy generates steam pressure within popcorn kernels; so, higher moisture results in faster expansion, but weakens the skin too much (called the "pericarp"), which causes failure (Anantheswaran and Lin, 1988).

Conventional microwave ovens (MWOs) operate on only one power level; either on or off. For example, when set at 60% power, a conventional microwave cooks at full power 60% of the time, and remains idle for the rest of the time. Inverter technology for true multiple-power levels (steady state) was recently conceptualized and developed by Matsushita Electric Industrial Company Limited in Japan, the company that owns the Panasonic brand. On the basis of comparative tests, they claim the following advantages in nutritional retention:

- Vitamin C (in cabbage)—72% versus 41%
- Calcium (in cabbage)—79% versus 63%
- Vitamin B1 (in pork)—73% versus 31%

Foods cooked in their inverter microwave over a conventional MWO, respectively. It would be interesting to see an unbiased, statistically designed, experiment conducted to confirm these findings.

Panasonic press release
"Now You're Cooking—Revolutionary microwave technology from Panasonic", May 2, 2002

The coded regression models that we obtained around this time are

$$\hat{y}_1 = 66.50 - 10.25A - 8.50B - 10.75AB$$
$$\text{Log}_{10}\,\hat{y}_2 = 0.023 - 0.16A - 0.31B$$

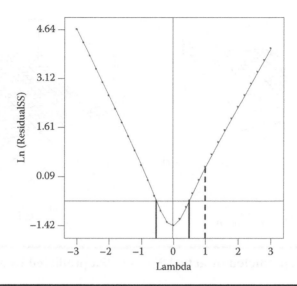

Figure 6.7   Box–Cox plot for UPK predicted via a linear model.

Notice that the base-ten logarithm is applied to the second response (UPK). This is a bit tricky to uncover, but it becomes obvious if you do a Box–Cox plot (see Figure 6.7) on the linear model (no interaction term).

The R^2_{Pred} for this transformed model is 0.89 versus 0.93 for the alternative 2FI model in original units:

$$\hat{y}_2 = 1.45 - 0.55A - 0.90B + 0.40AB$$

We used this when going over the same data in *DOE Simplified*. The models for UPK (y_2)—actual (no transformation) 2FI versus linear in log scale (transformed)—look different, but they produce similar response surface graphs as you can see in Figure 6.8a and b.

As George Box says, "All models are wrong, but some are useful." This is a common case, in which various models will do almost equally well for approximating the true response surface. As Box says, "Only God knows the true model," so, we will arbitrarily use the log-linear model (pictured in Figure 6.8b) as the basis for predicting UPK in the multiple-response optimization, even though the 2FI model may make more sense.

The objectives are obviously to minimize percent UPK and maximize taste. However, just as obviously, it will be impossible to achieve 100% yield (zero UPK) of the product. (Maybe not impossible, but very likely you'd burn up the popcorn, the microwave, and possibly the entire building in the process!) We must build in some slack in both objectives in the form of

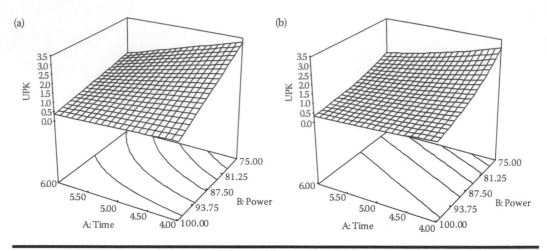

(a)

(b)

Figure 6.8 (a) UPK predicted by actual-2FI. (b) UPK predicted by a log-linear model.

minimally acceptable thresholds. Here's what we propose for limits in the two responses measured for the popcorn experiment:

1. *Taste.* Must be at least 65, but preferably maximized to the upper limit of 100. (Recall that the original taste scale went from 1 to 10, this being multiplied by 10 to make it easier to enter the averaged results without going to decimal places.)
2. *UPK.* Cannot exceed 1.2 ounces, but preferably minimized to the lower theoretical limit of 0.

CONSIDERATIONS FOR SETTING DESIRABILITY LIMITS

We made things easy by setting a stretch target of 100, the theoretical maximum, for the upper limit on taste. However, if this study were being done for commercial purposes, it would pay to invest in preference testing to avoid making completely arbitrary decisions for quantifying the thresholds. For example, sophisticated sensory evaluation may reveal that although experts can discern improvements beyond a certain level, the typical consumer cannot tell the difference. In such a case, perhaps 100 would represent the utmost rating of a highly trained taster, but the high limit would be set at 90, beyond which any further improvement would produce no commercial value—that is, increases in sales. Perhaps, a less-arbitrary example might be the measurement of clarity in a product such as beer. Very likely a laser turbidimeter might still register something in a beer that all humans would say is absolutely clear. In such a case, setting the low limit for clarity at zero on the turbidimeter would be overkill for that particular response. By backing off to the human limit of detection, you might create some slack that would be valuable for making tradeoffs on other responses (such as being less filling or tasting greater!).

These thresholds of a minimum of 65 and a maximum of 1.2 are translated via the models for taste and UPK (using the 2FI alternative), respectively, to boundaries on the graphical optimization plot shown in Figure 6.9. This frames the sweet spots, within one of which we now hope to identify the most desirable combination of time (factor A) and power (factor B).

With only two factors, it's easy to frame the sweet spots, because all factors can be shown on one plot. However, with three or more factors, all but two of them must be fixed, but at what level? This will be revealed via numerical desirability optimization; so, as a general rule, do this first and then finish up with graphical optimization. Otherwise, you will be looking for the proverbial needle in a haystack. We're really simplifying things at this stage by restricting this case study to two factors and putting the graphical ahead of the numerical optimization.

Table 6.2 specifies two combinations of processing factors that meet the requirements for taste and UPK, which we uncovered with the aid of computer software (Design-Expert) that features desirability analysis for numerical optimization of multiple responses.

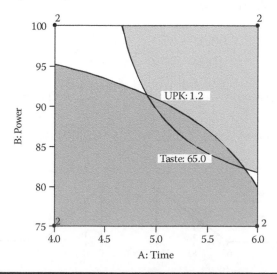

Figure 6.9   Sweet spots for making microwave popcorn.

Table 6.2   Best Settings for Microwave Popcorn

Rank	A: Time (minutes)	B: Power (percent)	y_1: Taste (rating)	y_2: UPK (ounces)	D
1	4	100	79.0	0.7	0.41
2	6	80.9	66.4	1.16	0.03

The last column shows the overall desirability D for each of the local optimums, the second of which is not easily found because it falls within a relatively small region (the unshaded one at the lower right of Figure 6.9).

The number-one ranked result does fairly well in terms of individual desirability for both taste and UPK as illustrated in Figure 6.10a and b. The smaller numbers, 32–81 for taste and 0.3–3.5 for UPK, along the base-lines are provided as reference points for the actual (observed) ranges of response.

The computations underlying D are

$$d_1 = \left(\frac{79 - 65}{100 - 65} \right) = \left(\frac{14}{35} \right) = 0.40$$

$$d_2 = \left(\frac{1.2 - 0.7}{1.2 - 0} \right) = \left(\frac{0.5}{1.2} \right) = 0.42$$

$$D = (0.40 * 0.42)^{1/2} = \sqrt{0.17} = 0.41$$

The second-ranked setup (time of 6 minutes at the power of 80.9%) produces barely desirable results:

$$d_1 = \left(\frac{66.4 - 65}{100 - 65} \right) = \left(\frac{1.4}{35} \right) = 0.04$$

$$d_2 = \left(\frac{1.2 - 1.16}{1.2 - 0} \right) = \left(\frac{0.04}{1.2} \right) = 0.025$$

$$D = (0.04 * 0.025)^{1/2} = \sqrt{0.001} \cong 0.03$$

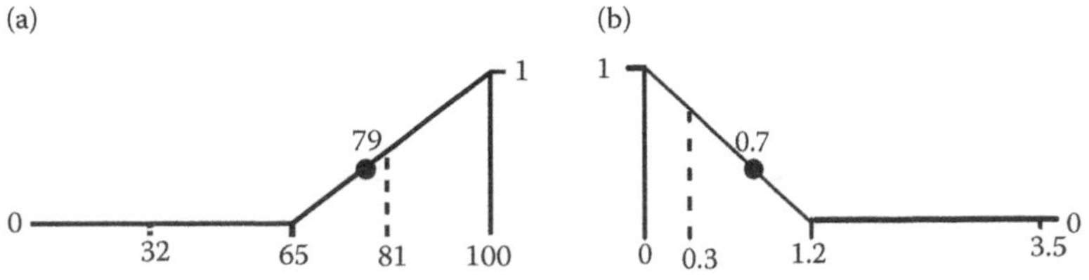

Figure 6.10   (a) Desirability ramp for taste. (b) Desirability ramp for UPK.

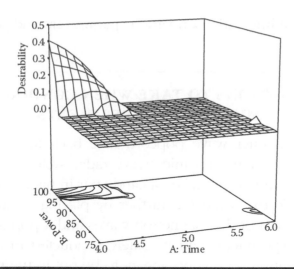

Figure 6.11   3D view of overall desirability.

It's amazing that this second optimum can be detected because it's just a tiny dimple in the zero plane as you can see in the 3D rendering of the overall desirability (D) in Figure 6.11.

To add insult to injury, the inferior setup for making microwave popcorn requires a substantially longer cooking time, 6 minutes versus 4 for the top-ranked solution (Figure 6.12a). Why not establish an additional goal to minimize time as shown in Figure 6.12b?

When the delicious aroma of popcorn permeates the atmosphere, an extra 2 minutes seems like an eternity! More importantly, reducing the time lessens the amount of electricity consumed—MWOs requiring a lot of power compared to most other home appliances.

We could refine the optimization further by playing with the parameters of weights and importance, but let's quit. When you start working with software offering these features for desirability analysis, feel free to do some

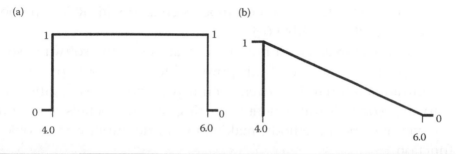

Figure 6.12   (a) Time kept within a range. (b) Time with the minimum goal.

trial-and-error experimentation on various parameters and see how this affects the outcome.

PRECAUTIONS TO TAKE WHEN CONDUCTING MICROWAVE EXPERIMENTS

Experiments with microwave popcorn date back to 1945 when Percy Spencer, while working on a microwave radar system, noticed that the waves from the magnetron melted a candy bar in his pocket! (Do not try this at home.) He experimented further by placing a bag of uncooked popcorn in the path of the magnetron waves and it popped.

Nowadays, experiments can be safely conducted in MWOs, but be careful when heating objects not meant to be put in the microwave. For example, one of the authors (by now you should not be surprised that this was Mark) bought a roast-beef sandwich that the fast-food restaurant wrapped in a shiny, paper-like wrapper. Knowing that it probably would not be a good idea to microwave, but wanting to see what would happen, he put it in the oven for reheating. The lightning storm that ensued was quite entertaining, and fortunately it did not damage the machine.

If you teach physics and want to demonstrate the power of microwaves more safely, see the referenced article by Hosack et al. (2002).

PRACTICE PROBLEMS

6.1 From the website for the program associated with this book, open the software tutorial titled "*Multifactor RSM*".pdf (* signifies other characters in the file name) and page forward to Part 2, which covers optimization. This is a direct follow-up to the tutorial we directed you to in Problem 4.1, so, be sure to do that one first. The second part of the RSM tutorial, which you should do now, demonstrates the use of software for optimization based on models created with RSM. The objectives are detailed in Table 6.3.

This tutorial exercises a number of features in the software; so, if you plan to make use of this powerful tool, do not bypass the optimization tutorial. However, even if you end up using different software, you will still benefit by poring over the details of optimization via numerical methods, making use of desirability as an objective function.

Table 6.3 Multiple-Response Optimization Criteria for a Chemical Process (Problem 4.1)

Response	Goal	Lower Limit	Upper Limit	Importance
Conversion	Maximize	80	100	+++
Activity	Target 63	60	66	+++

6.2 General Electric researchers (Stanard, 2002) present an inspirational application of multiple-response optimization to the manufacturing of plastic. The details are sketchy due to proprietary reasons; so, we've done our best to fill in gaps and make this into a case-study problem. It's just an example; so, do not get hung up on the particulars. The important thing is that these tools of RSM do get used by world-class manufactures such as General Electric Company.

The GE experimenters manipulated the following five factors:

A. Size

B. Cross-linking

C. Loading

D. Molecular weight

E. Graft

They did this via a CCD with a half-fractional core (2^{5-1}), including six CPs. Assume that at each of the resulting 32 runs, two parts were produced and tested for

1. Melt index (a measure of flowability based on how many grams of material pass through a standard orifice at standard pressure over a 10-minute period)

2. Viscosity

3. Gloss

On the basis of the models published by GE (with one curve, literally, thrown at you on the index to make the problem more interesting), we simulated the results shown in Table 6.4.

Note that levels were reported by GE only in coded format—again for reasons of propriety. Analyze these data (file name: "6-2 Prob— GE plastic.*") and apply numerical optimization to determine the optimum setup of factors in terms of overall desirability for the multiple responses. The criteria for individual desirabilities are spelled out in Table 6.5.

Table 6.4 Simulated Data for GE Plastic

Std	A: Size	B: X-Link	C: Load	D: Mol. wt	E: Graft	Melt Index	Visc.	Gloss
1	−1	−1	−1	−1	1	15.77	157	76.05
2	1	−1	−1	−1	−1	15.94	175.5	85.83
3	−1	1	−1	−1	−1	14.63	175.8	71.93
4	1	1	−1	−1	1	15.48	155.8	81.28
5	−1	−1	1	−1	−1	11.02	349.4	80.80
6	1	−1	1	−1	1	13.99	65.0	86.54
7	−1	1	1	−1	1	11.01	71.8	75.44
8	1	1	1	−1	−1	13.64	353.7	81.51
9	−1	−1	−1	1	−1	15.19	170.8	76.09
10	1	−1	−1	1	1	15.18	174.1	85.29
11	−1	1	−1	1	1	14.72	162.4	71.80
12	1	1	−1	1	−1	14.83	181.4	81.80
13	−1	−1	1	1	1	10.48	87.7	79.76
14	1	−1	1	1	−1	13.89	346.1	86.83
15	−1	1	1	1	−1	10.77	357.6	74.96
16	1	1	1	1	1	12.93	70.0	81.91
17	−2	0	0	0	0	11.25	197.3	71.80
18	2	0	0	0	0	14.67	196.9	87.57
19	0	−2	0	0	0	12.73	189.1	84.27
20	0	2	0	0	0	13.03	189	74.62
21	0	0	−2	0	0	15.46	151.2	78.61
22	0	0	2	0	0	9.442	240.6	81.57
23	0	0	0	−2	0	16.45	191.7	79.72
24	0	0	0	2	0	16.88	187.8	79.49
25	0	0	0	0	−2	12.67	348.3	79.8
26	0	0	0	0	2	13.32	51.2	79.82
27	0	0	0	0	0	13.72	195.7	79.66

(Continued)

Table 6.4 (*Continued*)   Simulated Data for GE Plastic

Std	A: Size	B: X-Link	C: Load	D: Mol. wt	E: Graft	Melt Index	Visc.	Gloss
28	0	0	0	0	0	11.73	183.3	80.29
29	0	0	0	0	0	12.29	200.5	80.36
30	0	0	0	0	0	12.33	186.4	79.73
31	0	0	0	0	0	11.27	184.5	80.8
32	0	0	0	0	0	13.8	166.7	79.76

Table 6.5   Multiple-Response Optimization Criteria

Response	Goal	Lower Limit	Upper Limit	Importance
Melt index	Target 14	13	15	+++
Viscosity	Target 160	120	190	+++
Gloss	Minimize	72	84	++

Appendix 6A: Simplex Optimization

Here, we provide some details on an algorithm called "simplex optimization," enough to give you a picture of what happens inside the black box of computer-aided numerical optimization. This algorithm performs well in conjunction with desirability functions for dealing with multiple responses from RSM experiments. It is a relatively efficient approach that's very robust to case variations.

A simplex is a geometric figure having a number of vertices equal to one more than the number of independent factors. For two factors, the simplex forms a triangle as shown in Figure 6A.1.

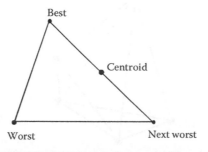

Figure 6A.1   Simplex with the evaluation of response.

The labels describe hypothetical rankings of overall desirability in the 2D factor space. Here are the rules according to the originators of simplex optimization (Spendley et al., 1962):

1. Construct the first simplex about a random starting point. The size can vary, but a reasonable side length is 10% of the coded factor range.
2. Compute the desirability at each vertex.
3. Initially, reflect away from the worst vertex (W). Then, on all subsequent steps, move away from the next-to-worst vertex (N), from the previous simplex.
4. Continue until the change in desirability becomes very small. It requires only function evaluations, not derivatives.

Figure 6A.2 may help a bit by picturing a move made from an initial simplex for three factors—called a "tetrahedron" in geometric parlance.

The use of next to worst (N) from the previous simplex ensures that you won't get stuck in a flip-flop (i.e., imagine going down rapidly in a rubber raft and getting stuck in a trough). The last N becomes the new W, which might better be termed as a "wastebasket" because it may not be worst. If you're a bit lost by all this, don't worry—as you can see in Figure 6A.3, it works!

This hypothetical example with two factors, temperature and catalyst, shows how in a fairly small number of moves, the simplex optimization converges on the optimum point. It will continue going around and around the peak if you don't stop it (never fear, the programmers won't allow an infinite loop like this to happen!).

So far, we've shown simplexes of a fixed size, but it's better to allow them to expand in favorable directions and, conversely, contract when things get worse instead of better. Another few lines of logic, which we won't get into,

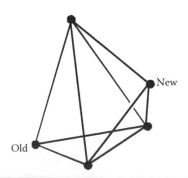

Figure 6A.2   Move made to a new simplex (tetrahedron) for a three-factor space.

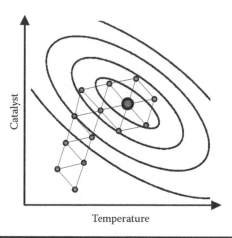

Figure 6A.3 Simplex optimization at work.

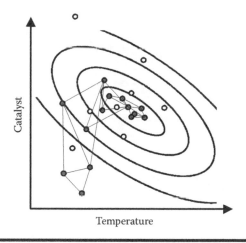

Figure 6A.4 Variable-sized simplex optimization at work.

accomplish this mission of making variable-sized simplexes (Nelder and Mead, 1965). Figure 6A.4 gives you a view of how this works for optimizing a response.

As you can see, the variable simplex feeds on gradients. Like a living organism, it grows up hills and shrinks around the peaks. It's kind of a wild thing, but little harm can occur as long as it stays caged in the computer. On the other hand, the fixed-size simplex (shown earlier) plods robotically in appropriate vectors and then cartwheels around the peak. It's better to use the variable-sized simplex for more precise results in a similar number of moves.

Chapter 7

Computer-Generated Optimal Designs

A man's got to know his limitations.

Clint Eastwood
From the movie Magnum Force

Ideally, you will find one of the two tried-and-true templates for RSMs, the CCD, or BBD, well suited to your experimental needs. However, these standard RSM designs may not conform to multifactor-operating constraints. For example, as shown in Figure 7.1, parts made in a hypothetical injection-molding machine come out defective when temperature and pressure fall outside of the specified zone.

The pressure needed to fill an injection mold depends on the viscosity of the molten plastic—the higher the temperature, the less viscous the plastic. If not done right, the process fails, either by underfilling the mold (short shot) or by putting in too much, creating parts with a flash. (A good example of applying DOE to flash reduction can be seen in a report from Lin and Chananda, 2003—two students at Kettering University who successfully applied the tools of DOE to make better replicas of their school mascot, the "Bulldog.")

As illustrated in Figure 7.1, multifactor linear constraints (MLCs) must be imposed to prevent bad things from happening to the plastic parts at certain

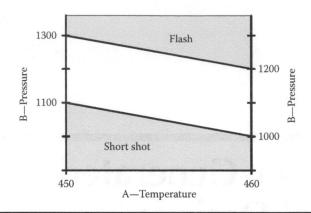

Figure 7.1   Constraints on a hypothetical molding process.

combinations of the two key processing factors. When both temperature and pressure are

- Set at low levels, they cause a short shot of plastic into the mold.
- Set at high levels, they cause an excess of material, which creates a flash along the edge of the part.

"FLASH" BACK TO PLAYING WITH PLASTIC ARMY MEN (AND OTHER TOYS)

Like the authors, if you grew up in the 1950s, little toy soldiers molded in soft plastic lurked about the play areas around your neighborhood. Almost every boy had bundles of these two-to-three-inch figures known as army men. They looked very lifelike except for the annoying ridges of plastic flash left over from the molding process. Owing to this manufacturing defect, some of these green-plastic army men looked more like the finned-back *Creature from the Black Lagoon* than a man in the military. The creators of the pioneering animated movie *Toy Story* (Pixar/Disney, 1995) used computer graphics imagery to animate an army of plastic soldiers, thus introducing a whole new generation to the joys of these tiny toys.

Tuesday's plastic corrosion awareness meeting was a big success.

Woody, the main character in *Toy Story*

Dealing with these multifactor constraints is our main focus for applying computer-generated designs.

OPTIMALITY FROM A TO Z

In Appendix 2A, we detailed the equation **X′X**, which produces the FIM. The FIM forms the core of the regression modeling at the heart of RSM. Over the years, various individual optimal characteristics of FIM have been detailed by statisticians and cataloged alphabetically. These criteria are very specific on their aims in regard to what they achieve for a test matrix. For example, let's consider a popular D-optimal criterion, one of the most popular for DOEs. In mathematical terms, the D-optimal algorithm maximizes the determinant of the FIM. The D stands for a determinant. A statistician seeing this definition would immediately realize and appreciate that this criterion minimizes the volume of the confidence ellipsoid for the coefficients. For those of you readers who work better with pictures than with matrix algebra, we detailed in Appendix 2A, an example of a confidence ellipsoid as shown in Figure 7.2.

The smaller the area of this ellipsoid, the more precisely one can estimate all coefficients (the β's) in the chosen model.

Many programs that provide tools for RSM now favor another alphabetic criterion labeled "I" over the D criterion. For a detailed comparison on these two criteria from a practical perspective, see Anderson and Whitcomb (2014, posted at www.statease.com/pubs/practical-aspects-for-designing-statistically-optimal-experiments.pdf).

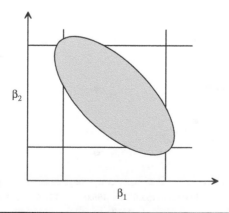

Figure 7.2 Example of a confidence ellipsoid.

Quantifying Operating Boundaries with Multilinear Constraints

As shown by the example pictured in Figure 7.1, simple upper and lower bounds on individual factors may not adequately define the experimental space. Additional constraints on linear combinations of factors must be imposed. Mathematically, this is expressed as

$$C_j \leq \beta_{1j}x_1 + \beta_{2j}x_2 + \cdots + \beta_{qj}x_q \leq D_j \quad j = 1, 2, \ldots, h$$

where the β_{ij} are scalar constants (these may be negative or zero) for up to q terms in the linear equation; and C_j and D_j are the lower and upper limit, respectively. There may be many such constraints—up to h of them, as noted to the right of the equation shown above.

For example, the MLC on making good plastic parts is

$$5600 \leq 10A + B \leq 5800$$

where A and B represent temperature and pressure, respectively. (To derive an MLC equation like this requires some mathematical trickery, which we will detail later in Appendix 7A.) A standard RSM design, such as a CCD or BBD, will not honor such complex constraints, thus placing points in regions that may be dangerous or simply impossible. This creates circumstances where it will be advantageous to make use of an algorithmically based design called "optimal." Figure 7.3 shows an optimal (D-criterion) selection

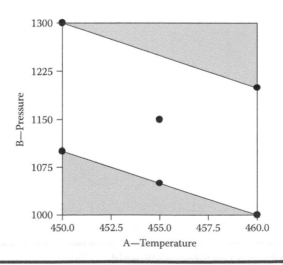

Figure 7.3   Computer-generated design for the molding process.

of points for fitting a quadratic model to temperature and pressure data kept within the MLC.

A major drawback to using optimal designs is that you must guess at the outset the exact model expected to best fit your data. Because the experiment needs to be conducted to determine the model, this creates a bit of a conundrum. As a general rule, the quadratic model, a second-degree polynomial, will be a good choice for RSMs. However, depending on the situation, you might go to a higher- or lower-order model or something in-between, for example, a 2FI equation.

CANDIDATE POINTS: A MATTER OF DEBATE

Typically, computer-based algorithms for an optimal design first generate candidate points and then use a stepping and exchanging process to select the experimental points. (See Appendix 7B for details. See also the note later in this chapter on "Candidate-Free Approaches for Exact Optimal Designs" for a heads up on an approach called "coordinate exchange.") The result may not be statistically optimal, but it will be nearly so, hopefully close enough for all practical purposes. How close the design gets to being truly optimal greatly depends on the quality of the candidate point set. As a general rule, the higher the degree demanded of the model, the more points should be considered for the design. For example, for a linear model, the extreme combinations or vertices in geometric terms, will suffice as a candidate set; but to fit quadratic terms, the midpoints will be needed, etc. (More details on how to generate a good candidate set are provided in Appendix 7B.)

Of course, the quantity of candidate points must exceed what's needed for fitting the specified model, which at the minimum equates to the number of terms. In the molding case, which involved two factors expected to exhibit quadratic behavior, six points were needed at a minimum because the desired equation includes six terms—the intercept plus coefficients for A and B (the main effects), the AB interaction, plus A^2 and B^2 (for modeling curvature). As a practical matter, the candidate sets cannot be too large because this will lead to time-consuming computations with diminishing returns in the optimality. At some stage, the process of selection and quest for absolute optimality must be terminated so that the experimenter can get to work.

> They have terminated jobs. They have terminated growth. And they
> have terminated dreams. And it's time to terminate them.
>
> **Arnold Schwarzenegger**
> *Points made by the former Governor of California and movie actor,*
> *while running his campaign in 2003*

To the straight eye, the design shown in Figure 7.3 is an odd guy. For
example, doesn't it seem a bit bizarre that the computer put a point at the
midpoint of the bottom edge but not on the top? Actually, either point,
or a midpoint on the left or right edge, would do. The choice is arbitrary.
Shouldn't some of the points be replicated to get a measure of pure error?
What if the quadratic polynomial does not adequately model the true
surface—will we be able to see that it does not fit? Maybe, we should take a
step back and think about what is really needed for a good RSM design.

Computer-Generated Design for Constrained Region of Interest

Here is a checklist (condensed from Box, 1982) against which we can gauge
the design quality, regardless of how factors will be constrained. Your design
should ideally

- Generate information throughout the region of interest
- Ensure that the fitted value, $\hat{y}$, be as close as possible to the true
  value
- Provide an internal estimate of error
- Give a good detectability of the lack of fit
- Allow experiments to be conducted in blocks
- Allow designs of an increasing order to be built up sequentially
- Be insensitive to wild observations
- Behave well when errors occur in the settings of factors (the x's)
- Not require an impractically large number of factor (x) levels
- Require minimal runs

The specific features common to all good designs (as defined by Box)
include

- Even coverage inside and around the surface of experimental (x) space
- Replicates for what Box calls an "internal estimate of error" (also known as pure error)
- Excess unique design points for testing the lack of fit

Figure 7.4 shows a new design for injection molding, which we constructed with these wish lists in mind.

How does this selection of points stack up against that chosen by the computer on a strictly D-optimal criterion (Figure 7.3)? Notice that we replicated a number of points as indicated by 2's. This provides a measure of pure error. We also added several new and unique points for testing the lack of fit. Filling in some spaces, such as the middle of the upper constraint, provides checkpoints on the model, assumed by the computer (for D-optimal purposes) to be quadratic, but in reality perhaps more complex—cubic or higher.

It cannot be emphasized enough that design optimality, whether it be D or some other flavor of the alphabet soup, depends on what you assume to be the model. As Box said in his 1982 article, "…the problem of experimental design is full of scientific arbitrariness. No two investigators would choose the same variables, start their experiments in the same place, change variables over the same region, and so on…we…make the problem worse… by introducing arbitrariness for purely mathematical reasons" (p. 51).

To illustrate how arbitrary decisions on the model lead to a nonsensical design, let's work through a simple problem that typically comes up in the

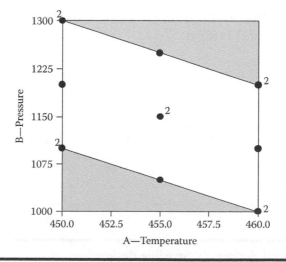

Figure 7.4 Modified design for the molding process.

career of a technical professional. Imagine that you've been given a mission to determine how a process works. Assume that, without a doubt, it will produce a response, y, as a linear function of an input, x. Expect that results will vary at any given setting of x. You are allowed only two runs within a constrained range. The possibilities are infinite between the low and high limits of x; so, let's simplify matters by establishing a candidate set of points at regular intervals of one-quarter, which in coded factor space will be −1, −0.5, 0, +0.5, and +1. Obviously, you will locate the two runs at the extremes of −1 to +1. If the response surface is linear, why bother putting design points in the interior? For example, if you put the two points at a more closely spaced interval such as −0.5 to +0.5, the variation in y would create a much bigger impact on the fit—the line would wiggle all over the place. The basic idea of optimality is to select design points that will generate the strongest model. In statistical terms, this results from a subset of candidate points that minimizes the overall joint CI illustrated in the note titled "Optimality from A to Z...."

Now, what if you cannot say without a doubt that the true response surface is linear? Let's allow for some flexibility in setting factor levels by providing not just two, but eight runs to work with. The design chosen by an optimal criterion is shown in Figure 7.5a. It's not very imaginative, but you ought to expect something like this from a computer because it does not waver in its goal of fitting a linear model. We, as human beings, are very capable of doubting assumptions. Therefore, in our sensible design in Figure 7.5b, we put half of the runs in the interior region, thus allowing for the possibility of curvature in the response. This alternative design is no longer optimal statistically, but it's a much more sensible way to spend the eight runs in the experimental budget.

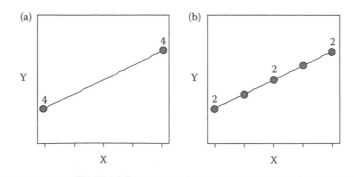

Figure 7.5   (a) Optimal design. (b) Sensible design.

GETTING THE COMPUTER TO BE MORE OPEN-MINDED ABOUT WHAT'S OPTIMAL

Here's how we suggest you to build and then augment optimal designs so that they make more sense for experimental purposes:

1. Using an optimal criterion, choose p points from within the experimental region, where p represents the number of coefficients in the chosen model, including the intercept. (See details on how to do this in Appendix 7C of this chapter.)
2. To provide one element for the lack-of-fit test, add at least three new unique checkpoints using the distance criterion to fill the holes in the feasible design region. (This criterion chooses a candidate point, not already included in the design by optimal selection, for which the minimum Euclidean distance to the existing points is at a maximum. It goes on from there until it adds; however, many checkpoints are desired.)
3. The lack-of-fit test requires a measure of pure error; so, we recommend you to select at least four points already chosen in steps 1 and 2 for replication. A good statistical criterion for choosing replicate points is leverage. Points with the highest leverage are the ones that you want to replicate.

To live up to the title of this book, which promises to simplify complex topics, we will relegate the mathematical details and the political debates on alphabetic flavors for optimality to the statistics professors and their graduate students. Let's see if we can get by with only a light KISS (keep it simple, statistically) that stimulates an appreciation for this computer-based design tool.

A Real-Life Application for Computer-Generated Optimal Design

Andy Scott, an investigator in the Pharmaceutical Development Sciences Department at GlaxoSmithKline's (GSK) Research and Development Center in the United Kingdom, asked the authors for advice on how to deal with a complex constraint on two air pressures, designated as A and B. The process, the purposes of which were not disclosed by GSK, required differential pressures to maintain a flow of air in only one direction. The constraints given by Andy are detailed below

- A from 5 to 8
- B from 7 to 9.5
- B must always be at least +1 greater than A

The last requirement is a multilinear constraint that can be mathematically expressed as $1 \leq B–A$. This was an ideal application for an optimal design. (At the time, this experiment was conducted and the D criterion reigned.) Aided by software, Andy laid out the experiment as shown in Figure 7.6. It's geared to fit a quadratic model, but he augmented the base design (the big, black circles) with replicate points for the estimation of pure error (the ones designated by 2) plus additional unique runs to test for the lack of fit (small, open circles). This latter set of runs, called "checkpoints," filled in open spaces. (See the note titled "Getting the Computer to be More Open-Minded…" for details on our criterion for the replicates and checkpoints.)

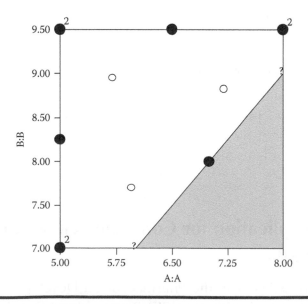

Figure 7.6   Computer-generated optimal design (modified) for an air pressure experiment.

In Figure 7.6, notice that only one run falls along the diagonal-operating constraint. If this boundary defines an area of interest, it would be easy to add the vertices along this edge at the (A,B) coordinates of (6, 7) and (8, 9). These two points (identified with question marks) were included in the candidate point set (see Appendix 7B) but were not chosen by the computer based on the optimal criterion.

The lesson here is that there's nothing sacred about this optimal (D) or any other computer-generated design. However, as the number of factors increases to three and beyond, it becomes very difficult to visualize the design space. Then, you'd better have a good software program for laying out a sensible design according to the 10-point checklist provided earlier in this chapter.

Other Applications for Optimal Designs

Our main focus for applying computer-generated optimal designs has been to deal with complex constraints that make standard RSM options infeasible. However, the optimal designs will also be useful for other circumstances that are not amenable to a standard RSM:

- *Specification of nonquadratic polynomial predictive models.* For example, cubic terms may be required based on prior process knowledge. In this case, neither the CCD nor the BBD would provide enough data for the complete third-order model.
- *Categorical factors.* For example, a variety of catalyst types and configurations must be tested at varying numerical process conditions such as time, temperature, pressure, and flow rate. In this case, two categorical factors (several catalyst types made into a variety of catalyst configurations) must somehow be combined with the numerical factors.

If neither of these scenarios applies to your experimental problem, and you need not contend with complex constraints like that in the molding case, we then advise you to steer clear of computer-generated optimal designs and choose a standard RSM design such as a CCD. It's been shown that a standard CCD is more robust than D and other alphabetic–optimal criteria, for example, I is optimal, in the likely event that only some of the terms in the quadratic model turn out to be significant (Borkowski and Valeroso, 2001). Also, by using tried-and-true textbook designs, you avoid the odd layouts generated by the computer using black-box algorithms for optimality.

PRACTICE PROBLEMS

7.1 After a series of screening and in-depth factorial designs, the search for a process optimum has been narrowed down to two factors, ranging as follows:

A. 110–180

B. 17–23

However, it's been discovered that due to equipment limitations, the combination of factors A and B must not fall below the constraint depicted in Figure 7.7.

Notice that, at the lowest level of A, factor B must be at least 19. At the lowest level of B factor, A must be at least 180. To further complicate matters, the experimenters suspect that the response surface may be wavy; in other words, the standard quadratic model used for RSM may fall short of providing the accurate predictions. Therefore, a cubic model is recommended for the design.

To learn how to do all this with the software provided, follow the tutorial on the website for the program associated with this book titled "*Multifactor RSM*".pdf (* signifies other characters in the file name) found via the "Optimal Numeric" link. But first, you ought to try to convert the complex constraint (represented by the diagonal line in Figure 7.7) into an equation. Hint: see Appendix 7A.

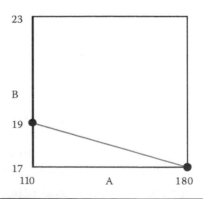

Figure 7.7   Constraint for Problem 7.1.

7.2 This is a real-life example using a manufacturer's equipment specifications. No response data will be provided—we will consider only the
DOEs. A processor of high-tech plastic sought optimum levels for two
key factors affecting the operation of an extruder:
   A. Screw speed, 300–600 revolutions per minute (rpm)
   B. Throughput, 600–1400 pounds per hour (pph)
   They desired the highest-possible throughput of the product
exhibiting a maximum flex modulus (a measure of stiffness in
pounds per square inch [psi]). However, due to torque limitations,
only a limited amount of material, 600 pph, can be processed
through the extruder when it runs at the lower screw speed (300
rpm). Only at the higher screw speed of 500 rpm can the maximum
desired output (1400 pph) be attained. It will not pay to break the
machine; so, we must derive a multilinear equation to quantify the
operating constraint. Unless you are really adept at math, even a
simple specification like this will take some effort to translate; so,
we will do this one for you. (Someone has to do the math because
the constraints must be put in a specific format for entry into the
software.)

$$B \le 600 + \left((A - 300) \left(\frac{1400 - 600}{500 - 300} \right) \right)$$

$$B \le 600 + \left((A - 300) \left(\frac{800}{200} \right) \right)$$

$$B \le 600 + 4A - 1200$$

$$B \le 4A - 600$$

   Using the provided software, set up an RSM design geared for a
quadratic model (hint: choose "Optimal (custom)" and then click
the Edit Constraints button to enter the multilinear constraint as
$B-4A \le -600$). Evaluate the resulting design and consider whether you
could do any better.

Appendix 7A: How to Quantify Multilinear Constraints

Consider the experimental region depicted in Figure 7A.1.

It represents three factors that can individually range as follows:

A. 1000–1700
B. 175–225
C. 0.2–0.7

But if all are set simultaneously high, the process exceeds its operating limits. Thus, we've imposed a complex constraint defined by the triangle at the upper-right corner of the cube with vertices (A, B, and C) or (1400, 208, and 0.3). Consider this to be a barrier beyond which the experimenter must not venture.

For purposes of quantifying this MLC, let's first define a cuboidal geometry by rescaling the three factors from 0 to 1, rather than the standard coded levels of −1 to +1. Before going any further, let's define some terms:

- LL is the lower limit—coded 0 (e.g., $LL_A = 1000$)
- UL is the upper limit—coded 1 ($UL_A = 1700$)
- CP is the constraint point for any given factor ($CP_A = 1400$)

In coded space, each edge of the hypercube will be one unit in length (0–1). Now, by applying one of the following two equations, we will determine what proportion of the unit length falls within the feasible region:

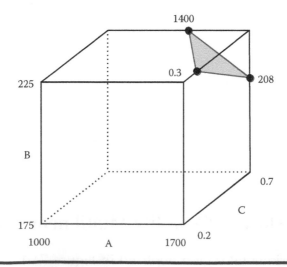

Figure 7A.1   Complex constraint.

- $[x - LL/CP - LL]$ if you want the given factor x to exceed its constraint point
- $[UL - x/UL - CP]$ if you want the given factor x to stay below the constraint

Whichever equation applies, it must be equal to or greater than one. For our example, we've specified the latter case for all three factors; so, the multilinear constraint is

$$1 \leq \frac{1700 - A}{1700 - 1400} + \frac{225 - B}{225 - 208} + \frac{0.7 - C}{0.7 - 0.3}$$

The last step is to simplify the function of factors so that it conforms to the linear format $\beta_1 A + \beta_2 B + \beta_3 C \ldots$. This can be done in our example problem by some arithmetic:

$$1 \leq \frac{1700 - A}{300} + \frac{225 - B}{17} + \frac{0.7 - C}{0.4}$$

$$(300)(17)(0.4)(1) \leq 17(0.4)(1700 - A) + 300(0.4)(225 - B) + 300(17)(0.7 - C)$$

$$2040 \leq 11560 - 6.8A + 27000 - 120B + 3570 - 5100C$$

$$-40090 \leq -6.8A - 120B - 5100C$$

Appendix 7B: How to Generate a Good Candidate Set

We suggest that to generate a good candidate set of points for optimal selection, you characterize the geometry of your experimental region (the x-space) as follows:

1. Vertices—the corners of the design space
2. Centers of edges—midpoints between adjacent vertices
3. Thirds of edges—two points equally spaced between adjacent vertices
4. Triple blends—averages of three adjacent vertices
5. Constraint plane centroids—center points in the planar surfaces of the experimental region (the convex hull)
6. Checkpoints—average of the centroid and vertices
7. Interior points—average of the centroid and (if the points are selected) centers of edges, thirds of edges, and constraint plane centroids
8. Overall centroid—center of the design space

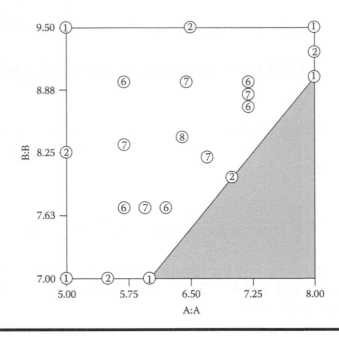

Figure 7B.1   Candidate set for an air-flow study.

Which of these geometric points should be selected as candidates depends on the model you want to fit with your RSM design:

A. Linear: vertices, checkpoints, and the centroid
B. 2FI model: same as linear
C. Quadratic: same as linear and 2FI models plus centers of edges, constraint plane centroids, and interior points
D. Cubic: same as quadratic plus thirds of edges and triple blends

Figure 7B.1 shows the candidate set for the air-flow study, which was designed for a quadratic model. The points are keyed by a number to the list provided earlier on the x-spaced geometry.

Appendix 7C: An Algorithm for Optimal Point Selection

Here is an algorithm we've used to find an approximately optimal design from candidate point sets chosen in the manner described in Appendix 7B:

1. Select a nonsingular initial design of p points via a random process.
2. Start exchanging points in the initial design. Begin by adding one point that increases the optimal criterion the most. Exchange it with the point that decreases it the least. Continue one-point exchange until there is

no improvement. Then, perform two-point exchange steps (two for two), three point, and so on. If there is improvement at any stage, start over with one-point exchanges.

As an option to the procedure outlined in step 1, you can get started by making a random selection of points from the candidate set. The entire process can be repeated from any number of random bootstraps. However, be forewarned: If the candidate list is large and/or the degree of the design is large, this optimal point selection process may take a long time even on very powerful computers. Obviously, the computations will become much more intense as you increase the number of points exchanged in step 2. For example, going up to 10-point exchanges multiplies the time required to develop the design by about five.

CANDIDATE-FREE APPROACHES FOR EXACT OPTIMAL DESIGNS

Candidate sets grow exponentially by the number of factors and by degree of the polynomial that the experimenter wants to fit with the data. Eventually, this process becomes so unwieldy that the process must be brought to a halt before achieving anywhere near optimal results. A candidate-free algorithm called "coordinate exchange" (Meyer and Nachtsheim, 1995) reduces computational time for experiments on 10 or more factors. For the air-flow case discussed earlier, Figure 7C.1 shows

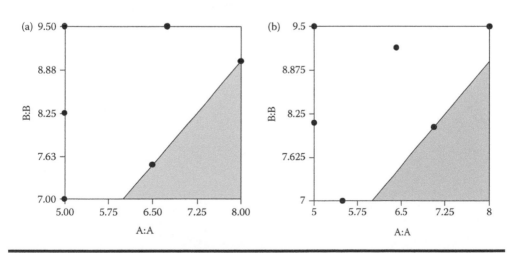

Figure 7C.1 (a, b) D versus I—optimal designs via candidate-free coordinate exchange.

which combinations of factors the coordinate exchange algorithm picks by D versus I (now favored for RSM) optimal criteria.

Compare the coordinate selection of six points in Figure 7C.1a (D-optimal design at the left) to the six solid-black ones shown in Figure 7.6 picked from a candidate set (also based on D-optimality). Notice that the newer design puts an additional point along the diagonal-operating constraint. Also, the determinant $\mathbf{X'X}$ (FIM with variance fixed at 1) for quadratic modeling is 42.7 for the optimal design done via coordinate exchange, which compares favorably against the FIM of 41.0 achieved via the more conventional methods (requiring generation of a candidate set).

P.S. Coordinate exchange can be applied to the distance-based criterion as well, thus freeing up the dependency on candidate sets when adding checkpoints to test for the lack of fit.

Chapter 8

Everything You Should Know about CCDs (but Dare Not Ask!)

Theory guides, experiment decides.

I.M. Kolthoff
*University of Minnesota professor known as
the father of analytical chemistry*

From the very beginning (Box and Wilson, 1951), statisticians have debated criteria for the ideal structure of the CCD, in particular, the number of CPs per block and how far out the axial levels (star points) should go. In this chapter, we will discuss what's good in theory versus more practical considerations for real-life experimentation.

Let's start by tackling the toughest issue for experimenters wanting to do RSM via CCDs—where to place their star points. For over half a century, the dogma has been that stars must go outside of the factorial box as shown in Figure 8.1.

Going to these extremes offers advantages, the most obvious of which is the greater leverage for estimating the main effects and curvature. The main disadvantage of venturing outside the box is equally obvious—the stars may break the envelope of what's safe or even physically possible. For example, what if you include a factor such as a dimension of a physical part and the lower star point comes out negative? Or, worse yet, the equipment may fail catastrophically when certain operating limits are exceeded at the upper star

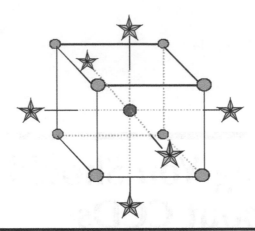

Figure 8.1 Standard CCD with stars outside of the box.

points. There's a more subtle disadvantage to the standard construction for the CCD; experimenters often find it inconvenient to adjust their process to the five levels required of each factor:

■ A low axial (a star at the lowest level)
■ A low factorial
■ A CP
■ A high factorial
■ A high axial (a star at the highest level)

Therefore, as we mentioned briefly in Chapter 4, the experimenter may opt to do an FCD (see Figure 4.2).

Now that you know you don't *have* to go to the extremes shown in Figure 8.1, it may be easier to remain open-minded about the possibilities. Let's get into the statistical details on why Box and Wilson were so keen for you to do this and determine exactly where they wanted to put the stars (and why). When constructed properly, the CCD provides a solid foundation for generating a response surface map by providing the necessary runs to fit a quadratic polynomial. Figure 8.2 lays out the geometry for a generic CCD on two factors.

A CCD always contains twice as many star points as there are factors in the design. The axial distances to the star points are designated by the Greek symbol alpha (α). They are measured in terms of coded units, where plus-or-minus 1 represents the factorial ranges. Typically, the alpha value varies from 1 (the face-centered option) to the square root of the number of factors ($\sqrt{k}$), which produces a spherical geometry. However, even higher values may be chosen. For example, Box and Draper (1987, pp. 305–311)

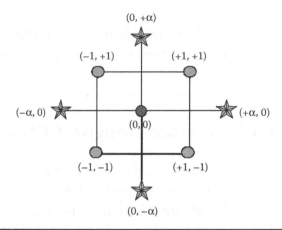

Figure 8.2   Coded levels for CCD.

show a CCD for three factors with the star points placed at plus-or-minus 2 (>√3 = 1.73). Myers et al. (2016, p. 395) also present a three-factor case, but they put the stars to 1.682 units from the CP. Oddly enough, in the CCD example detailed in Chapter 4, the coded alpha value of 1.682 was used to generate the star points for the three factors (see Table 4.1 for the actual levels). Why use such an odd value: 1.682? It turns out that at this particular placement of the star points, a CCD with three factors exhibits a feature much coveted by statisticians—rotatability.

THE FATHER OF QUADRATIC EQUATIONS

Diophantus was an Alexandrian (Egypt) whom many consider to be the father of algebra. He is best known for his text *Arithmetica*, which presented the solution for quadratic equations—the primary model for RSM. Historians know very little about his life or even when he lived other than it was in the early centuries A.D. Even his epitaph is a puzzle (Newman, 1956), but if you are adept with fractions, it reveals Diophantus's age at death:

> …God vouchsafed that he should be a boy for the sixth part of his life; when a twelfth was added, his cheeks acquired a beard; He kindled for him the light of marriage after a seventh, and in the fifth year after his marriage He granted him a son. Alas, late-begotten and miserable child, when he [Diophantus's son] had reached the measure of half his father's life, the chill grave took him. After consoling his grief by this science of numbers for four years, he [Diophantus] reached the end of his life.

Can you figure out how long Diophantus lived? (Hint: determine the common denominator for the three fractions of his life before getting married.)

Rotatable Central Composite Designs (CCDs)

Rotatability—what's in it for you? Well, first of all, recall from Chapter 4 that the standard error plot (Figure 4.9a and b) for the design we held up as a sterling RSM exhibited perfectly circular contours, which indicates that equally precise predictions will be obtained at any location equally distant from the CP of the experimental space. Any other pattern would indicate that the design favors moving from your bull's eye at the middle in one direction versus another, thus indicating a bias on the part of the experimenter. As you've learned by now, bias is a four-letter word in statistical jargon, one that must be avoided if at all possible.

Consider an example of a nonrotatable design—a full three-level factorial. The standard error plot for a 3^2 design (nine runs, only one at a CP) can be seen in Figure 8.3.

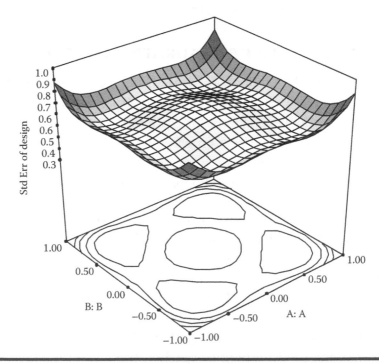

Figure 8.3   Standard error plot for a three-level design (nonrotatable).

First of all, notice that the projected contours are not circular, thus indicating nonrotatability. However, as a more practical matter, you see that there are four well-predicted pockets at regions that probably hold no more interest for the experimenter than any other combination of factors.

ROUND AND ROUND ON ROTATABILITY: SOME DIZZYING DETAILS

Box and Hunter (1957) were the first DOE experts to tout rotatability as a desirable feature for RSM. To generate this property in a CCD, the value of alpha must conform to the following function:

$$\alpha = [2^{(k-p)}]^{1/4}$$

Recall that 2^{k-p} specifies a two-level design on k factors, where minus p represents the fraction. In other words, for a CCD to be rotatable, the coded axial distance of its star points must equate to the fourth root of the number of experimental runs in the two-level factorial portion. Designs conforming to this specification will generally place the factorial points and star points on different spheres. For example, a three-factor CCD puts stars at plus-or-minus 1.68 coded units $(=[2^3]^{1/4} = 8^{1/4})$ versus a radius for cube points of 1.73. Thus, this design, like most rotatable CCDs, is not equiradial.

So long as we're going around in circles in this sidebar, it's worth noting, while on the topic of equiradial geometry that this does describe a class of rotatable RSM designs for two factors, the most popular of which is the hexagon, featuring six evenly spaced points in a circle, plus a number of CPs. The minimum-run equiradial design for fitting a quadratic polynomial is the pentagon, but this should be avoided due to having points with a leverage of 1. (Perhaps, such a dangerous design should be left to experimenters working on behalf of the U.S. Defense Department, who make their home in the famous Washington, DC building of the same geometry.) A special case of an equiradial design is the octagon, which doubles as the two-factor rotatable CCD; so, it gets used quite often by practitioners of RSM.

A much better alternative is the two-factor CCD pictured in Figure 8.4 with a rotatable alpha value of 1.414 and four replicates of the CP (five in all). Notice in Figure 8.4 how the standard error for prediction, plotted to the same scale as before, now exhibits symmetry.

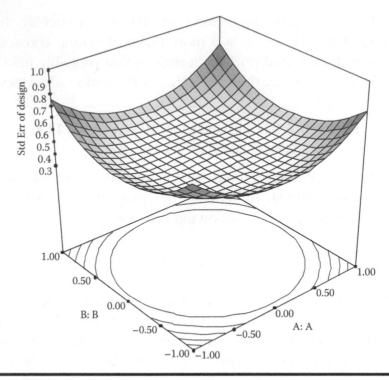

Figure 8.4   Standard error plot for a two-factor rotatable CCD.

Also, by bulking up on the CPs, the error for response prediction is minimized in the bulls-eye region and it increases only gradually until you get very near the factorial boundaries (±1 coded unit from the center). We advise that you be wary when wandering outside of this box because of the rapid increase in standard error, and dangers of extrapolating into space that's only been "scouted" by the venturesome stars. However, as suggested in Chapter 4's not "Going Outside the Box," after running a rotatable CCD, you will do well by moving beyond the factorial limits to the standard error limit set by these points.

Going Cuboidal When Your Region of Operability and Interest Coincide

Here is a communication (Anderson, 2001) with an actual experimenter that raises some practical questions about where to place star points in a CCD:

> I am having a tough time trying to decide between two CCDs, one with alpha of 1.682 [rotatable] and the other with alpha of 1 [face-centered]. I have three factors, each of which can vary from

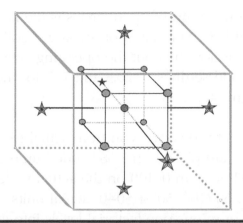

Figure 8.5 Rotatable CCD fit into a cubical region of operability.

0 to 100. Negative values are inadmissible. Therefore, I generated
the first [rotatable] design by setting factor ranges in terms of alpha.

In this case, the experimenter's region of operability (0–100) and the
region of interest coincide. Imagine trying to force fit the rotatable three-
factor CCD (coded alpha = 1.682) into a cubical cage with each side of
100 actual units in length. The factorial cube is now pushed well within the
middle of the region of interest (see Figure 8.5).

Clearly, only a minor portion of the region of operation/interest will be
explored by the RSM design, specifically $(1/1.682)^3$ or 0.21. In other words,
nearly 80% of the region of interest is left relatively unexplored.

Now, we see that the experimenter's idea of doing an FCD makes
much more sense. By looking at a two-factor projection of the 3D region,
Figure 8.6 provides a comparative view of the FCD versus CCD inscribed
within the operating limits of 0–100.

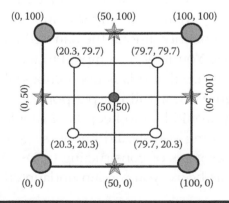

Figure 8.6 FCD versus CCD.

The smaller square made up of open circles is the factorial region abandoned from the rotatable CCD in favor of the larger, solid-pointed factorial part of the FCD that now fills out the operating envelope. You may be wondering about the odd settings of 20.3 and 79.7 for the CCD. A bit of arithmetic solves this mystery:

1. Start by working out the conversion from actual to coded units. To be rotatable, the stars are placed on 1.682 coded units that are out from the (50, 50) CP. They span 0–100 on the actual scale. Thus, 1.682 coded units convert into 50 (100–50 or 50–0) actual units.
2. Convert the plus-or-minus one factorial levels into actual ranges: 50/1.682 = 29.7.
3. Add and subtract this value from the actual coordinates at the CP to get the values of 79.7 (=50 + 29.7) and 20.3 (=50 – 29.7).

This bother is alleviated by the use of software that will set up CCDs, but that still leaves the inconvenient settings to be run on the process—all the more reason to abandon this design in favor of the FCD.

Of course, everyone knows that you cannot put a square peg in a round hole, or in this case, expect a cube to fill a sphere. By changing alpha to 1, you pull the axial points into the face of the cube, which creates a design that's no longer rotatable (spherical standard error contours) but rather cuboidal, as can be seen in Figure 8.7, which depicts the standard error (plotted to the same scale as the previous plots) for the FCD discussed above (pictured in Figure 8.6).

In conclusion, for five or fewer factors, we advise you to make use of an FCD, rather than a rotatable CCD, when you want to explore the entire region of process operability. As we will detail later in a sidebar called "Dampening Inflation of Variance," at larger numbers of factors (>5), FCDs exhibit an excessive collinearity among the squared terms, which can be quantified via their VIFs. You may be tempted to try a BBD for situations like this. However, be aware that this is not a cuboidal design because, as we illustrated in Figure 5.1, it will not put points at the corners of your space. (Sorry, we really would like to keep RSM simple, but at least, we waited until this "everything you should know" stage to delve into all the pros and cons of various ways of constructing CCDs. We must assume that if you're still reading this book, you remain undaunted and dare ask about such things.)

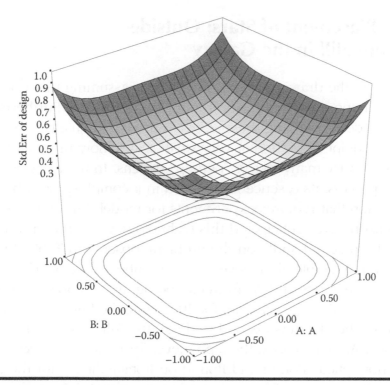

Figure 8.7 Cuboidal standard error for FCD.

CUBOIDAL SPACES FOR THE OFFICE
(A TIME-OUT FROM STAT STUFF)

If you work for a large corporation that keeps you cooped up in a cubicle, take a break from all this humdrum and rent or stream the movie *Office Space*, written and directed by Mike Judge, Twentieth Century Fox (1999). To avoid the closed-in feeling of myriad cuboidal spaces, one high-tech Californian client of the authors eliminated all walls and outlawed all opaque furnishings. It featured mesh-backed chairs and desks made out of wire. Perhaps, cubes aren't that bad after all!

We don't have a lot of time on this earth! We weren't meant to spend it this way. Human beings were not meant to sit in little cubicles staring at computer screens all day… filling out useless forms… and listening to eight different bosses drone on about mission statements.

Spoken by one of the oppressed workers in *Office Space*

Practical Placement of Stars: Outside the Box but Still in the Galaxy

Let's go back to the drawing board on CCDs by assuming that operating limits do not impose constraints on the region of interest. Furthermore, we ought to consider expanding these RSMs to more than just two or three factors. Robotic equipment, often at very small scales, now makes it feasible to simultaneously vary many factors, perhaps dozens. In many cases, there is no physical process: its essence is captured in a complex, black-box computer simulation that requires experiments for model development. Surely, Box and Wilson never anticipated this technology when they invented the CCD in 1951. To get a handle on the problem, observe in Table 8.1 what happens to the rotatable alpha level as the number of factors (k) escalates.

To keep the number of runs from exploding, we've applied a fraction (p) to the core factorials. For example, for 10 factors, the fractional factorial is 2^{-3} or 1/8th of the 512 runs needed to do all the full number of two-level combinations. As p increases, the alpha level for rotatability decreases (see the earlier note titled "Round and Round on Rotatability…" for the formula), but even so, it remains so far out that one must question its practicality. Myers et al. (2016, p. 406) propose setting alpha at the square root of the number of factors ($\sqrt{k}$) to create a spherical design such as that shown in Figure 8.8.

However, as you can see in Table 8.1, this spherical specification for the star points achieves KISS (keeping it simple, statistically), but does not make them any more practical.

We propose a much more radical paring of the alpha level for star points: fourth root of the number of factors ($k^{1/4}$). The results of this calculation can be seen in the last column of Table 8.1. As discussed in the sidebar "Dampening Inflation of Variance," this design specification offers a good tradeoff of practicality versus statistical properties.

Table 8.1   Impact of Increasing Factors on Star-Point Placement

k	p	Rot. α	$\alpha = k^{1/2}$	$\alpha = k^{1/4}$
2	0	1.414	1.414	1.189
5	1	2.000	2.236	1.495
10	3	3.364	3.162	1.778
20	11	4.757	4.472	2.115

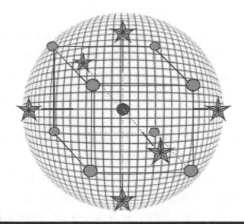

Figure 8.8   Spherical CCD.

DAMPENING INFLATION OF VARIANCE

As stars in a CCD get pulled into the face of the factorial hypercube, the VIFs for the quadratic terms (x_i^2) increase, thus making it harder to estimate their model coefficients. Fortunately, the VIFs are dampened considerably by only a small increase in alpha levels above the face-centered value of 1. This is illustrated in Figure 8.9, which shows how VIF varies as a function

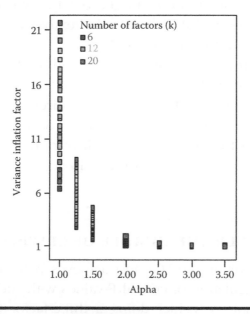

Figure 8.9   VIFs as a function of alpha and k.

of alpha and k—the number of factors from 6 to 20, in some cases with more than one result shown due to options on how far to fractionate the core factorial. As a rough rule of thumb, the VIF at alpha 1.00 equals the number of factors. Although it's not that relevant to our conclusions, this should help you to interpret the shades of gray keyed to k.

Here's another rule of thumb: VIFs above 10 indicate an unacceptably high multicollinearity (Montgomery, Peck, and Vining, p. 119). On this basis, it can be inferred from the graph that an FCD ($\alpha = 1.00$) will not be a good choice unless dictated by operating constraints. Our initial thought for KISS was to fix alpha at a level of 1.50 where VIFs fall well below the level of 10. But as a refinement to counteract the higher VIFs for designs with many factors ($k \geq 6$), we suggest setting alpha on the basis of the fourth root of k, if practical, but not less than 1.5.

For more details, see Whitcomb and Anderson (2003), posted at www .statease.com/publications/case-studies/practical-versus-statistical-aspects-of-alternating-central-composite-designs.html.

CPs in Central Composite Designs

Our clients often ask if it would be alright to cut down or possibly even eliminate the replication of CPs called for by textbooks and associated software for CCDs. In short, the answer is "No!"

Recall from the previous chapter that a good RSM design should provide for testing of LOF. Otherwise, you cannot tell how well your model fits the actual response data. This statistic requires a measure of pure error, which comes only from true replication. To achieve any sort of reasonable power for the LOF test, three df of pure error are mandatory. If only CPs are replicated, then, you must do at least four of them to get over this threshold.

GETTING OVER THE HURDLE OF CRITICAL F-VALUES

With fewer than four df, the LOF test achieves very little power, in part due to the rapid escalation of critical F-values with such meager samples of data. For example, consider doing a three-factor CCD. For uniform precision, we suggest that this design includes six CPs, but you may be tempted to do fewer. Think again after seeing these critical F-values at the 5% risk level as a function of the number of "CPs":

CPs	df	$F_{(5,df)}$ for LOF
1	0	Not possible
2	1	230.2
3	2	19.30
4	3	9.01
5	4	6.26
6	5	5.05

To construct this table, we assumed a full quadratic model that leaves five df for a residual to test against many df of pure error for the F-test on LOF. Notice how it stabilizes at four df and beyond.

Furthermore, as we discuss in Appendix 8A at the end of this chapter, if the CCD is broken up into blocks and you desire them to be orthogonal, the number of CPs not only increases, but they must also be split into different design segments in a specific manner.

Despite all these admonitions, nonstatistical industrial experimenters generally persist in viewing the replication of CPs to be a waste of valuable time and resources. Let's look at this controversy from another angle—uniformity of precision for response predictions. As you've seen in Figures 8.4 and 8.7, a well-constructed CCD or FCD provides a relatively flat and low standard error throughout the middle of the factorial region. On the other hand, as you can see in Figure 8.10, cutting back on the number of CPs degrades this uniformly precise pattern of standard error (plotted to the same scale as before). It shows what happens if you cut back to only one CP from the five in the original two-factor rotatable CCD illustrated in Figure 8.4.

Flip back and forth between Figures 8.4 and 8.10. Enough said?

Final Comments: How We Recommend That You Construct CCDs

So now, you know a lot more about the construction of CCDs than you ever wanted to learn. Unless you are a statistician (who laps up this kind of stuff), it's likely that at this stage you suffer from TMI. Therefore, we decided that it would be merciful to try summarizing our recommendations on CCDs.

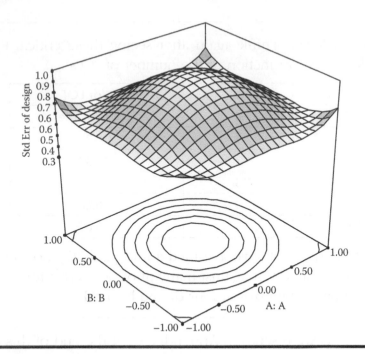

Figure 8.10   Standard error plot for a two-factor rotatable CCD with only one CP.

First of all, you must consider how your region of interest relates to the operating boundaries for the experimental factors. For example, consider the situation pictured in Figure 8.11.

It depicts an FCD (equivalent in this case to a full three-level design) for two factors held within operating limits. The circles around the CP show that it's replicated a number of times—we recommend five for a good test on LOF. In cases such as this, where the regions of interest and operation coincide, we recommend the FCD.

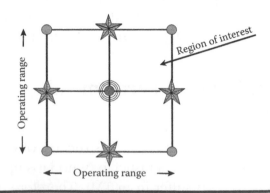

Figure 8.11   Region of interest versus operability.

On the other hand, when operating limits do not come into play relative to the factor ranges you wish to explore; in other words, you will not be pressing against the process envelope, the more traditional form of CCD, with star points outside the box, which will provide much improved predictive properties. However, to keep the range of these wayward stars within a realistic striking distance, we recommend setting them at the fourth root of the number of factors ($\alpha = k^{1/4}$). At this setting of alpha, the design will remain nearly rotatable but it will not waste nearly as much space as at the exactly rotatable level. For example, recall that the standard three-factor CCD ($\alpha = 1.6823$) restricted the ideal volume for prediction (inside the factorial box) to about 1/5th ($[1/1.6823]^3 = 0.21$) of the overall region spanned by the star points. At the lesser value of alpha that we recommend ($k = 3^{1/4} = 1.316$), this region of interest expands to nearly one-half ($[1/1.316]^3 = 0.44$) of the overall experimental volume.

The number of CPs in CCDs should never be less than five in our opinion. The more, the better, particularly as the number of factors increases. Ideally, sufficient replication of CPs can be done to achieve a relatively uniform precision inside the factorial ranges. However, although more research is needed in this area, it may take some multiple of the number of factors to accomplish this objective. For now, to KISS, we advise increasing the replication of CPs to 2 times the number of factors (2k), but capping the number at 10. So, for example, eight CPs in a four-factor CCD might be a practical number.

YET ANOTHER OPTION: ORTHOGONAL QUADRATIC CCDs

We cannot leave well enough alone on all the permutations for setting up CCDs; so, we ask you to consider one more: choosing the alpha level for the star points in such a manner that perfects the VIF for the quadratic terms (x_i^2). Achieving VIFs of 1 makes these model coefficients for estimating curvature orthogonal to the main effect and the 2FI term. For example, by setting alpha to 1.414 ($\sqrt{2}$) on a two-factor CCD, you make it rotatable, but the VIFs for A^2 and B^2 are 1.02, which is not quite perfect. If you adhere to the philosophy (paraphrasing George Box) "when doing something really worthwhile you may as well do it right," and you believe orthogonality is of utmost importance, then, you must set the alpha level at 1.2671 for a two-factor CCD with five CPs. We won't go into the details,[*] but by increasing the CPs on the design to eight, the alpha for orthogonality of the quadratic terms increases to 1.414 and then the design is also rotatable.

---

[*] This criterion for setting alpha was brought up originally by Box and Wilson in the landmark paper on RSM in 1951. If you really want to learn more on the topic, see their paper cited in the references for this book.

PRACTICE PROBLEMS

8.1 University of Minnesota food scientists (Richert et al., 1974) used a CCD to study the effects of five factors on whey protein concentrates, a popular diet supplement for body builders. Also, whey proteins, when properly processed, are beneficial in the development of foams characteristic of frozen desserts, whipped toppings, meringues, and mousses.

WHERE THERE'S A WHEY

Whey is the watery part (called the "serum") of milk that is separated from the curd in making cheese. Unless you work in the dairy industry (or engage in body building), your only familiarity with this term may be from the nursery rhyme:

> Little Miss Muffet, sat on a tuffet,
> Eating her curds and whey.
> Along came a spider,
> Who sat down beside her
> And frightened Miss Muffet away.

P.S. A "tuffet" is either a mound of earth or a three-legged stool. (Source: http://sircourtlynice.blogspot.com/2010/09/tuffets-and-how-to-sit-on-them.html.)

The factors, with ranges noted in terms of alpha (star levels), were

A. Heating temperature: 65–85 degree celsius per 30 minutes
B. pH level: 4–8
C. Redox potential: −0.025 to 0.375 volts
D. Sodium oxalate (Na "Ox"): 0–0.05 molar
E. Sodium lauryl sulfate (Na LS): 0%–0.2% of solids

There were nine responses, but we will look at only three

1. Whipping time in minutes (to produce a given amount of foam)
2. Time at the first drop in minutes (a measure of foam stability)
3. Undenatured protein in percent

Denature means to render unfit, therefore, the *un*denatured protein must be maximized. The whipping time should be minimized and the stability maximized.

The experimenters chose a rotatable CCD based on a one-half fraction for the cube portion (2^{5-1}). The data are shown in Table 8.2. Be careful when you set up this CCD—all points must fall within the specified ranges. If supported by your software (it is in the one provided with this book), enter the limits in terms of the alpha levels, which are the axial (star) points.

This is a popular design (e.g., it was also used by GE scientists to study plastic processing as presented in Problem 6.2). However, as we discussed in this chapter, rotatable CCDs do a poor job of exploring the x-space, that is, the region defined by the process limits. What would have been a good alternative for these food scientists studying the production of whey protein concentrates?

Develop predictive models for the three responses of interest. With these many factors (five), you may discover that many terms become insignificant in the ANOVA. Therefore, we suggest you to apply model reduction. We did this in Chapter 2 for the Longley data and discussed the pros and cons of it in the sidebar titled "Are You a 'Tosser' or a 'Keeper'?" in Chapter 4. Recall our advice that you be vigilant for cases where a factor and its dependents (e.g., A, AB, AC, and A^2) are all insignificant. Is it possible to model any or all of the whey protein responses with only a subset of the tested factors—in other words, one or more factors reduced out of the model? If so, this will simplify the response surfaces and keep things simpler all around.

Also, do not be shy about applying response transformations. By now, you've seen this done numerous times throughout the book to produce better predictive models.

Table 8.3 provides criteria for a multiple-response optimization of the process for making whey protein concentrate.

Can you find a sweet spot for the meringue and mousse lovers?

8.2 Leonard Lye, a professor of engineering and applied science at Memorial University of Newfoundland, contributed the following case study. It is based on the DOE Golfer, a machine he invented to teach RSM.

Table 8.2　Data for Whey Protein Study

Std	A: Heat (deg C)	B: pH	C: Redox (volt)	D: Na OX (Molar)	E: Na LS (percent)	Whip Time (minutes)	Time First Drop (minutes)	Protein (percent)
1	70	5	0.075	0.0125	0.15	4.75	4.5	80.6
2	80	5	0.075	0.0125	0.05	4.00	7.5	67.9
3	70	7	0.075	0.0125	0.05	5.00	8.3	83.1
4	80	7	0.075	0.0125	0.15	9.50	17.0	38.1
5	70	5	0.275	0.0125	0.05	4.00	6.7	79.7
6	80	5	0.275	0.0125	0.15	5.00	9.5	74.7
7	70	7	0.275	0.0125	0.15	3.00	12.0	71.2
8	80	7	0.275	0.0125	0.05	7.00	36.0	36.8
9	70	5	0.075	0.0375	0.05	5.25	4.0	81.7
10	80	5	0.075	0.0375	0.15	5.00	5.0	66.8
11	70	7	0.075	0.0375	0.15	3.00	12.5	73.0
12	80	7	0.075	0.0375	0.05	6.50	20.0	40.5
13	70	5	0.275	0.0375	0.15	3.25	15.0	74.9
14	80	5	0.275	0.0375	0.05	5.00	7.5	74.2
15	70	7	0.275	0.0375	0.05	2.75	18.5	63.5
16	80	7	0.275	0.0375	0.15	5.00	12.0	42.8

(Continued)

Table 8.2 (Continued)   Data for Whey Protein Study

Std	A: Heat (deg C)	B: pH	C: Redox (volt)	D: Na OX (Molar)	E: Na LS (percent)	Whip Time (minutes)	Time First Drop (minutes)	Protein (percent)
17	65	6	0.175	0.025	0.10	3.75	12.0	80.9
18	85	6	0.175	0.025	0.10	11.00	8.5	42.4
19	75	4	0.175	0.025	0.10	4.50	4.5	73.4
20	75	8	0.175	0.025	0.10	4.00	10.5	45
21	75	6	−0.025	0.025	0.10	5.00	9.0	66
22	75	6	0.375	0.025	0.10	3.75	9.0	71.7
23	75	6	0.175	0.000	0.10	3.75	9.0	77.5
24	75	6	0.175	0.050	0.10	4.75	10.0	76.3
25	75	6	0.175	0.025	0.00	4.00	16.0	67.4
26	75	6	0.175	0.025	0.20	3.50	8.5	86.5
27	75	6	0.175	0.025	0.10	4.00	11.0	77.4
28	75	6	0.175	0.025	0.10	3.50	9.0	74.6
29	75	6	0.175	0.025	0.10	3.50	9.0	79.8
30	75	6	0.175	0.025	0.10	4.00	10.0	78.3
31	75	6	0.175	0.025	0.10	3.50	9.5	74.8
32	75	6	0.175	0.025	0.10	3.00	11	80.9

Table 8.3   Multiple-Response Optimization Criteria for Whey Protein Concentrate

Response	Goal	Lower Limit	Upper Limit	Importance
Whip time	Minimize	1	4	++
Time first drop	Maximize	10	20	+++
Protein	Maximize	80	100	++++

(a)   (b)

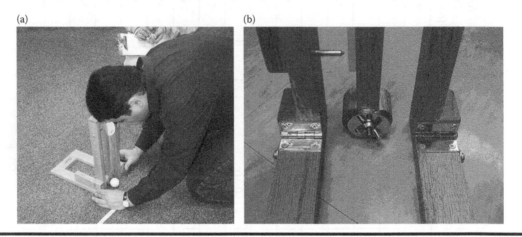

Figure 8.12   (a) Student setting up a golf machine. (b) Golf machine club head.

Figure 8.12a shows a student preparing a putt with the DOE Golfer. The weight of the club head can be adjusted by adding washers as shown in the close-up of the golfing machine in Figure 8.12b.

Many factors can be varied in the DOE Golfer, including length of the club, angle of swing, weight of the club, and the type of ball. Table 8.4 shows an experiment conducted on the golfing machine by a group of students who studied three of these four factors. They ended up doing an FCD in two blocks.

Notice that the first block is a full two-level factorial (2^3) with five CPs. What did the students see from the analysis of this block alone that led them to augment it with a second block of six face-centered star points (standard-order rows 14 through 19) plus two more CPs (20, 21)? (Hint: do a square root transformation on the response.) On the basis of the predictive model generated from this RSM, what would be a good setup for a six-foot (72-inch) putt with the DOE Golfer, assuming you can give or take 2 inches?

Table 8.4   FCD Done on DOE Golfer

Std	Blk	A: Length (inches)	B: Angle (degree)	C: Weight (washers)	Distance (inches)
1	1	5.5	20	0	2.2
2	1	11.5	20	0	17.3
3	1	5.5	60	0	16.5
4	1	11.5	60	0	82.3
5	1	5.5	20	2	8.2
6	1	11.5	20	2	23.2
7	1	5.5	60	2	37.0
8	1	11.5	60	2	115.3
9	1	8.5	40	1	40.0
10	1	8.5	40	1	37.0
11	1	8.5	40	1	38.5
12	1	8.5	40	1	34.5
13	1	8.5	40	1	39.5
14	2	5.5	40	1	16.0
15	2	11.5	40	1	55.0
16	2	8.5	20	1	14.0
17	2	8.5	60	1	71.0
18	2	8.5	40	0	31.0
19	2	8.5	40	2	41.5
20	2	8.5	40	1	35.5
21	2	8.5	40	1	36.0

BECOMING AN ACE AT DOING DOE

For bonus marks in his graduate DOE course, Professor Lye organized a golf tournament in the carpeted faculty lounge. Evidently, the grain of this "green" proved to be a decisive factor according to his report on the outcome:

The group that won the tournament clearly had the best design and was the most meticulous. They noticed that direction is important. Going north-to-south is different from going south-to-north, etc. The winners took this into account and that made quite a difference. They even got a hole-in-one for one hole and the maximum number of strokes they needed for any hole was two. It was amazing really.

Other groups needed more than 5 or 6 strokes for one hole. Their first stroke usually got them fairly close to the hole (maybe 2 inches away), but some groups completely forgot about this possibility and failed to develop a proper model for this short distance. Also, if they overshot with the first stroke, the next one had to come from the opposite direction. If the students did not consider direction as a factor, then their prediction was not accurate.

The winning group considered all these possibilities. The moral here is not to assume that the floor is flat or the carpet is uniform in all directions.

(For more details on the DOE Golfer and how you can buy one for educational purposes, see www.statease.com/golftoy.html.)

Appendix 8A: Some Details on Blocking CCDs and Issues of Orthogonality

One of the more practical features built into CCDs by Box and Wilson is their ability to be built up block by block. This can be likened to modular construction of prefab units for customized housing—build only what you really need to cover the required space. For example, the three-factor CCD described in Chapter 4 actually began with a block of 12 factorial runs detailed in Chapter 3. A second block of eight runs was required to make it an RSM design, thus allowing proper modeling of the curvature. As discussed, this three-factor CCD was rotatable. However, it comes up to just the tiniest bit short of achieving orthogonal blocking.

Technically, an RSM is said to be orthogonally blocked if it is divided into segments in such a manner that block effects do not affect the coefficient estimates for the second-order (quadratic) model. This can be quantified in terms of the correlation (r); a value of zero ($r = 0$) indicates orthogonality.

In this case, design evaluation reveals a 0.028 correlation between the block coefficient (included in the predictive model as a correction factor for changes from one time frame to the other) and each of the squared terms (A^2, B^2, and C^2). For what it's worth (not much!), by reducing the axial distance of the star points very slightly from the rotatable level of 1.68 to a value of 1.63, this troublesome (only to certain statisticians!) correlation could be eliminated (zeroed out). (Of course, this makes the resulting design imperceptibly nonrotatable.)

In general, for a CCD with k factors done in two blocks (factorial points in one and axial in the other) to be orthogonal, the star points must be set at

$$\alpha = \sqrt{k \left[\frac{1 + (n_{s0}/2k)}{1 + (n_{f0}/2^{k-p})} \right]}$$

where n_{s0} and n_{f0} are the number of CPs in the axial and factorial blocks, respectively. Since the terms 2k and 2^{k-p} represent the number of star points and factorial points, respectively, you can see that the orthogonal choice for alpha boils down to a function of the number of factors and the ratio of CPs to other points in each of the two blocks. Solving for the three-factor example we get

$$\alpha = \sqrt{3 \left[\frac{1 + (2/2 * 3)}{1 + (4/2^{3-0})} \right]} = \sqrt{3 \left[\frac{1 + (2/6)}{1 + (4/8)} \right]} = 1.632$$

In some cases, CCDs can be set up with an alpha value that achieves both rotatability and orthogonality (Ah…sleep at last for the RSM perfectionist!). For example, consider the four-factor design broken up into either two or three blocks as follows:

1. 16 (2^4) factorial points plus four CPs
2. 8 (2*4) star points plus another two CPs

   or

1. 8 (one-half) factorial points plus two CPs
2. 8 (the other half) factorial points plus two CPs
3. 8 (2*4) star points plus another two CPs

As noted earlier in a sidebar titled "Round and Round on Rotatability...," to be rotatable, alpha must be

$$\alpha = [2^{(k-p)}]^{1/4} = [2^4]^{1/4} = 2$$

To achieve orthogonal blocking

$$\alpha = \sqrt{4\left[\frac{1 + (2/2 * 4)}{1 + (4/2^{4-0})}\right]} = \sqrt{4\left[\frac{1 + (2/8)}{1 + (4/16)}\right]} = 2$$

This calculation was done for the two-block option that keeps the whole factorial core, but even if it gets split into half, thus producing a third block, the number of CPs to factorial points is maintained (4/16 versus 2/8 + 2/8). Therefore, the alpha value of 2 is orthogonal for either option (two versus three blocks).

An equation for simultaneous rotatability and orthogonality of blocked CCDs can be found in Box and Draper (1987, p. 510, Equation 15.3.7). They account for the possibility, which seems to be very unrealistic, given how big RSM designs can be, that an experimenter might fully replicate the factorial and/or star-point portions of the CCD. This has a bearing on the selection of alpha for rotatability and/or orthogonality.

For some values of k, it will be impossible to find a rotatable CCD that blocks orthogonally, but you can come very close on one or both measures with a given alpha value for the star points by a judicious selection of CPs. These specifications are already well documented in RSM textbooks (Box and Draper, p. 512) and statistical journals (Draper, 1982) that you can find encoded in off-the-shelf statistical software offering optimization tools of DOE (Design-Expert, Stat-Ease, Inc.), so, we will not enumerate them here.

IS CCD STILL ROTATABLE WHEN ONLY
LINEAR TERMS ARE SIGNIFICANT?

Ideally, you will at most need a full quadratic polynomial to adequately model the surface for any given response. However, the more things you measure from experimental runs, the more likely that some responses, especially those for which the testing cannot be done very precisely, will only need a linear model at the most. How does this affect issues of rotatability versus orthogonality? Fortunately, things simplify for the lower-degree polynomial: orthogonality for first-order models guarantees rotatability. Therefore, the standard two-level factorials (2^{k-p}) and other orthogonal screening designs are rotatable for estimating the main effects. All CCDs, regardless of the choice for alpha level (where you put the star points), are orthogonal to the main effects and thus rotatable for the first-order model. Isn't that nice? Sleep tight!

Appendix 8B: Small Composite Designs (SCDs) to Minimize Runs

In the jargon of RSM, the adjective "composite" describes an experimental design made up of discrete blocks geared to fit a quadratic model. Since Box and Wilson introduced these RSM designs in 1951, much work has been devoted to reduce the number of required runs. The minimum number of runs to fit a quadratic model with k factors equals the parameters p, which can be broken down by degree:

- Zero: 1 constant (β_0) often described as the intercept or mean
- First: k main-effect coefficients (β_i)—the slopes for each individual (i) factor
- Second:
    - k(k–1)/2 coefficients for 2FIs
    - k pure quadratic (squared terms) for curvature coefficients

This simplifies to

$$p = (k + 1)(k + 2)/2$$

For example, a minimum-run RSM on 10 factors requires 66 (=(10 + 1)(10 + 2)/2) unique combinations of the factors. Designs like this are considered to be saturated with the maximum number of factors in the given number of runs.

An experimenter may want a minimal-run composite design when runs are extremely expensive, difficult, or time consuming.

Early on, statisticians seeking minimal-run composite designs realized that it would be acceptable if the factorial core aliased the main effects with

2FIs, provided it did not alias these second-order terms with each other (see the sidebar titled "An Asterisk on Resolution" for details).

AN ASTERISK ON RESOLUTION

In 1959, Hartley suggested reducing the cube portion of composite designs to resolution III, provided no 2FIs are aliased with other 2FIs. This specific subset of fractional designs, from the standard 2^k family, is designated as resolution III*. It's OK to have the main effects aliased with 2FIs in the cube portion because the star points fill in the gap on the estimates of the main effects.

Successors (Westlake, 1965) came up with alternative cores based on irregular fractions of 2^k factorials, such as three-quarter replicates. However, perhaps, the most popular approach currently for minimal-run CCDs, developed by Draper and Lin (1990), makes use of Plackett–Burman designs for the two-level block of runs.

As shown in Table 8B.1, the Draper–Lin (D–L) small composite designs (SCDs) require far fewer runs than a full CCD, or even the ones with many factors that feature a fractional core. However, this reduction in runs comes at the cost of desirable design properties—balance, robustness, etc.

For a design with a better balance, consider another sort of SCD called the "hybrid" (Roquemore, 1976). These minimal-run (or nearly so) second-order

Table 8B.1   Number of Runs by Factor for Various Composite Designs

k	Min-Run	CCD $2^k/2^{k-p}$	MR5 CCD	SCD D–L	SCD Hybrid
2	6	13/–	–	–	–
3	10	20/–	–	15	11
4	15	30/–	–	21	16
5	21	50/32	–	26	–
6	28	86/50	40	33	28
7	36	152/88	50	41	46
8	45	282/90	60	51	–
9	55	540/156	70	61	–
10	66	–/158	82	71	–

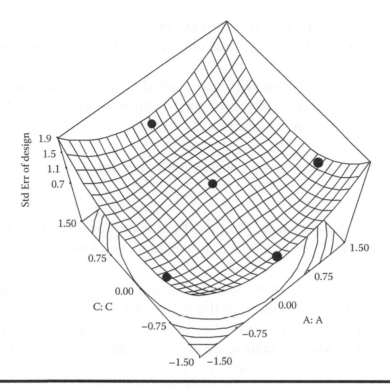

Figure 8B.1 Standard error plot for a hybrid design.

designs for 3, 4, 6, and 7 factors resemble a standard CCD for all but the last (kth) factor, which gets set according to model-optimality criterion. Montgomery et al. (2016, p. 458) say, "It has been our experience that hybrid designs are not used as much in industrial applications as they should be… because of the 'messy levels' required of the extra design variable." For example, notice the odd spread of points within the range of factor C on the standard error plot pictured in Figure 8B.1 for the three-factor hybrid design.

You'd best assign to this letter, or whichever comes last, the factor you find easiest to change in your process. Furthermore, we strongly advise that you add at least two additional CPs to hybrid designs—more for experiments on six or seven factors.

Templates for hybrid, D–L, and other types of SCDs are available via the Internet (Block and Mee). Many of them are encoded in RSM software such as Design-Expert from Stat-Ease, Inc.

Good Alternative to SCDs: Minimum-Run Resolution V CCDs

As a general rule, you get what you pay for with SCDs that are saturated with factors, or nearly so. Therefore, no matter which flavor of SCD you choose, D–L or hybrid, every run counts; so, these designs become extremely susceptible to outliers and outright failures that cause missing data.

We suggest a more-robust alternative called "Minimum-Run Resolution V [MR5] CCD" (Oehlert and Whitcomb, 2002), which as shown in Table 8B.1, requires fewer runs than standard CCDs, but does not shave so close to the bone as the SCDs.

The MR5 CCDs feature cores of irregularly fractionated two-level factorials that are

■ Equireplicated (the same number of plus and minus 1's for each factor)
■ Minimum-run (or plus 1 for equireplication) resolution V

They are provided for as many as 50 factors by the software accompanying this book. However, even with the minimal-run core, this 50-factor MR5 CCD requires 1382 runs. That may be doable, though, provided a robotic apparatus for the experimentation, or it being done only on a virtual basis, that is, via a computer simulation.

WHERE SHOULD STAR POINTS BE LOCATED FOR SMALLER-RUN CCDs?

Haven't you had enough of this yet? Fortunately, we need not open up another can of worms on how to set the alpha levels: Apply the same considerations for SCDs and MR5 CCDs as you would for any other CCD—that is, properties of rotatability, orthogonality of blocks, etc.

Chapter 9

RSM for Six Sigma

The native of one of our flat English counties ... retrospect of Switzerland was that, if its mountains could be thrown into its lakes, two nuisances would be got rid of at once.

Sir Francis Galton
Pioneer of Statistical Correlation and Regression

Now, we are ready to apply a tool called "POE" that applies the tools of calculus to find the flats on response surfaces. These regions are desirable, especially so if you are subjected to Six Sigma standards, because they do not get affected much by variations in factor settings. The idea is to seek out the high plateaus of product quality and process efficiency. Without the addition of POE as a criterion, computer-aided optimization might set your process on a sharp peak of response. Such a location will not be robust to variation transmitted from input variables.

**HOW OFTEN WOULD WORKERS BE TARDY
IF AT SIX SIGMA STANDARDS?**

In Chapter 1, we introduced this topic with a fun, but hopefully very relevant, study on how to reduce the variability of commuting time to work. For many workers, even those on salary, the boss exhibits little tolerance for deviation from the specified arrival time. The goal of Six Sigma is to reduce variation of processes, such as the commute to work, to such a degree that the failure rate drops to 3.4 parts per million or less. Let's see

---

if we can translate this statistic to something more meaningful. According to the Michigan Department of Career Development, manufacturers in the United States employ 2.3 million assemblers to put together automobiles, appliances, electronic products, and machines, as well as the related parts. Can you imagine out of this entire labor force that fewer than eight workers would be tardy on any given day? Work out the math for Six Sigma!

Developing POE at the Simplest Level of RSM: A One-Factor Experiment

POE is a scary subject because it requires the use of mathematical tools that may be rusting at the back of your mind from long disuse. Although you can employ software (e.g., Design-Expert from Stat-Ease, Inc.) to do the "heavy lifting," it may do you some good to do a light workout and tone up those math muscles. Fortunately, you've already developed a great deal of strength by learning how to make use of RSM for generating additive polynomial models that are extremely amenable to POE. For example, in Chapter 1, let's revisit the model that we fit to the data collected on drive time:

$$\hat{y} = 32.1 + 1.7x - 0.1x^2 + 0.00016x^3$$

This equation is a cubic polynomial, which as you observed in Figure 1.11, creates a wavy surface that contains both highs and lows. We want to locate these plateaus and valleys because they represent points where variations in the input factors (x) transmit the least to the response (y). They can be pinpointed via POE.

The first step is calculation of the response variance (σ_y^2)

$$\sigma_y^2 = (\delta_y/\delta_x)^2 \sigma_x^2 + \sigma_e^2$$

where σ_x^2 is the variance of the input factor x and σ_e^2 is the residual variance, which comes from the ANOVA. In the drive-time case, the standard deviation of Mark's departure time is estimated to be 5 minutes; so, the values in the equation for variance are

$$\sigma_x^2 = (5 \text{ minutes})^2 = 25$$

$$\sigma_e^2 = \text{Mean square residual from ANOVA} = 5.18$$

The partial derivative of the response y with respect to the factor x (δ_y/δ_x) for the drive-time model is

$$\delta_y/\delta_x = 1.7 - 0.2x + 0.00048x^2$$

The POE is conveniently expressed in the original units of measure (minutes of drive time in this case) by taking the square root of the variance of response Y via this final equation:

$$POE = \sqrt{\sigma_y^2} = \sqrt{[(\delta_y/\delta_x)^2 \sigma_x^2 + \sigma_c^2]}$$
$$= \sqrt{[(1.7 - 0.2x + 0.00048x^2)^2 * 25 + 5.18]}$$

From here, it becomes "plug and chug" for those who enjoy doing old-fashioned hand calculations. However, with the aid of software, the POE can be more easily (and accurately!) calculated, mapped, and visualized. Figure 9.1 shows the POE for the drive-time case study. It exhibits two local minima, one for the peak at 11.5 (a high flat point in Figure 1.11) and one for the valley at 31.5 (a low flat point). (We superimposed a lightly shaded facsimile of the predicted drive time from Figure 1.11 on the POE pictured in Figure 9.1 to help you make the comparison.)

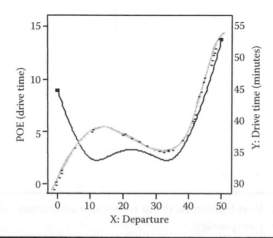

Figure 9.1 POE of drive time versus departure time.

Thus, POE finds two of the flats that are much desired by Six Sigma; so, life is good (so long as the commuter's tires do not go flat!). As noted earlier, to conserve sleep, the commuter (Mark) chose the second minimum POE (31.5 minutes past 6:30 a.m., or a bit after 7 a.m.). Obviously, this implies that he gets in later to work. It turns out that Mark has the luxury of adjusting a second factor: flexible hours for the work day.

The commuting case illustrates an ideal system where factors affect the response in various ways:

■ Nonlinear (departure time x)
■ Linear (adjusting the work hours within a daily flex-time window)

Then, to reduce variability, the experimenter simply adjusts the nonlinear factor to minimize POE, and as a necessary follow-up, resets the linear factor to get the response back on target.

WHY POE BECOMES IRRELEVANT FOR LINEAR MODELS

POE will be constant when the model is linear—it makes no difference where you set the factors: the variation transmitted does not change as shown in Figure 9.2b.

Therefore, linear factors can be freely adjusted to bring the response into specification while maintaining the gains made in reducing variation

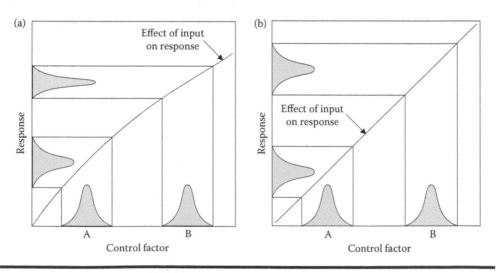

Figure 9.2 **(a) Variation transmitted via nonlinear responses. (b) Variation transmitted via linear responses.**

via POE on the nonlinear factors. In the case pictured above, take advantage of nonlinearity in response by setting the level of the first control factor (shown in Figure 9.2a) to reduce variation. Then, adjust the other factor (exhibiting the linear effect pictured in Figure 9.2b) to bring the response back to the desired set point.

Figure 9.3 shows the ideal outcome for the application of POE—a big reduction in process variability, with no deleterious impact on the average response.

With information gained from POE analysis, the experimenter can adjust the nonlinear factors to minimize process variation. This tightens the distribution as shown by the middle curve in Figure 9.3. However, the average response now falls upon the target. Therefore, a linear factor must be adjusted to bring the process back into specification (the "after" curve). Thus, the mission of a robust design and Six Sigma is accomplished.

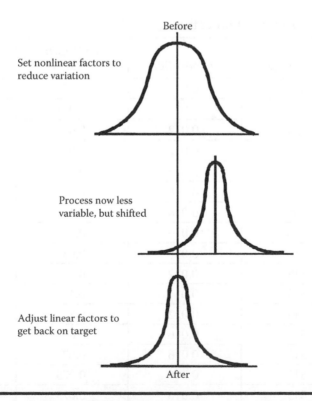

Figure 9.3   Before and after applying POE analysis.

An Application of POE for Three Factors Studied via a BBD

The drive-time example involves only one experimental factor. To get a better feel for the application of POE, let's look at a more complex case study, one that involves three factors on a highly automated lathe (Anderson and Whitcomb, 1996). The data from a BBD are shown in Table 9.1. The response, labeled "Delta," gives the deviation of the finished part's dimension from its nominal value in mils (0.001 inches). Ideally, this delta can be stabilized at, or near, zero.

Regression analysis reveals a significant ($p < 0.0001$) quadratic model, shown below in terms of actual factors:

$$y = 1.42 - 0.00172A - 98.4B - 25.0C + 0.235AB$$
$$- 0.0532AC + 2.25 * 10^{-6}A^2 + 406C^2$$

(*Note:* POE calculations, which we detail in Appendix 9A, operate on actual units of measure—not coded units.)

Table 9.1   Data from BBD on Lathe

Std	A: Speed (fpm)	B: Feed (ipr)	C: Depth (inches)	Y: Delta (mils)
1	330	0.01	0.075	0.0203
2	700	0.01	0.075	−0.357
3	330	0.022	0.075	−0.292
4	700	0.022	0.075	0.374
5	330	0.016	0.05	−0.394
6	700	0.016	0.05	0.227
7	330	0.016	0.1	0.572
8	700	0.016	0.1	0.208
9	515	0.01	0.05	−0.232
10	515	0.022	0.05	0.0886
11	515	0.01	0.1	0.123
12	515	0.022	0.1	0.471
13	515	0.016	0.075	−0.299
14	515	0.016	0.075	−0.183
15	515	0.016	0.075	−0.128
16	515	0.016	0.075	−0.146
17	515	0.016	0.075	−0.113

First, let's explore the response surfaces generated by this model to see whether parts can be lathed to the proper dimension (delta equal to zero). Only two factors can be chosen for any given contour (or 3D) plot. Rather than choosing these arbitrarily, make use of the perturbation plot shown in Figure 9.4.

Notice that factor C (the depth of a cut) makes the most dramatic impact on the response, whereas B (the feed rate in inches per revolution) looks linear. Factor A (speed in feet per minute) falls somewhere in-between. Therefore, a plot of A versus C will be the most interesting (Figure 9.5a and b). We will keep B at its midpoint.

The 3D plot in Figure 9.5b reveals a flat-bottomed valley—an ideal place to locate the process for minimal POE. On the 2D plot in Figure 9.5a, we highlighted the ideal contour at zero delta. It is nice to know that, at this mid-level feed rate, the lathe operators can choose any number of combinations for speed and depth to achieve their part specification. However, let's set the target aside for the moment and only concentrate on the calculation of the POE.

To compute POE, the variations in input factors must be specified as shown in Table 9.2. The relative variations are provided only for the sake of reference—to provide a feel on how well each factor can be controlled.

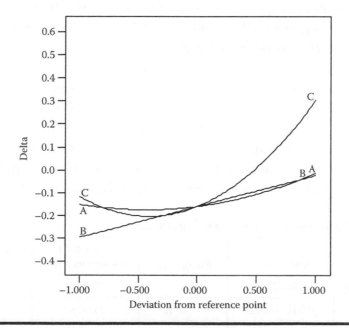

Figure 9.4 Perturbation plot of delta for the lathed part.

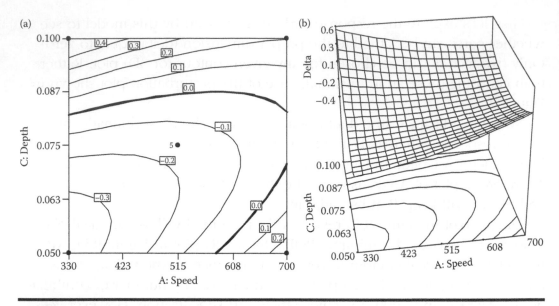

Figure 9.5 **(a) 2D contour plots of speed versus depth. (b) 3D surface for delta.**

For example, in this case, the depth (factor C) varied far more (25%) than speed (A—1.3%) on a relative basis. However, as one can infer from the broadly spaced contours, at certain locations, the response becomes insensitive to any variations in either of these two factors. To pinpoint this robust region, all we need to do is put the final piece of the puzzle into place—the standard deviation in response (provided from ANOVA by taking the square root of the mean square of the residual: 0.0675 mils). Then, the POE plot shown in Figure 9.6 can be generated.

It shows the most desirable location to be at the bottom end of both speed (A) and depth (C)—the lower-left quadrant in this experimental space. However, that cannot work because it will be put on the part that is off the target (i.e., delta will not be equal to zero). (Refer to Figure 9.5b to see this.) We must look out for a setting of factors that meets product specification at the minimum POE. This is just the sort of "less-filling/tastes great" multiple-response conflict that we resolved in Chapter 6 by the use of the desirability

Table 9.2 **Variation in Factors Affecting the Lathe**

Factor	Test Limits	Range (Δ)	Standard Deviation (σ)	Relative Variation (σ/Δ) (%)
A: Speed (fpm)	330–700	370	5	1.3
B: Feed (ipr)	0.010–0.022	0.012	0.003	25
C: Depth (in.)	0.05–0.10	0.05	0.0125	25

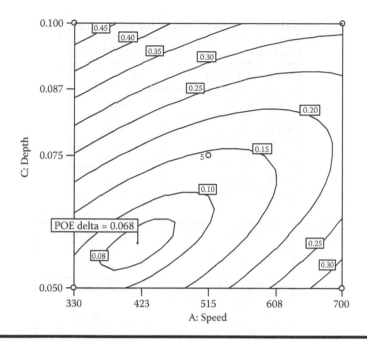

Figure 9.6   POE contour plot.

function. In this case, we set the maximum desirability for delta at zero and POE minimized as shown in Figure 9.7a and b.

The threshold limits, where desirability drops to zero, are set somewhat arbitrarily in this case, but some slack must be allowed to facilitate the tradeoff between the demands of multiple responses. Table 9.3 shows the outcome of a numerical optimization on the overall desirability. (The POE is actually treated as a second response, y_2.)

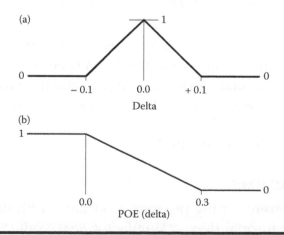

Figure 9.7   (a) Desirability ramp for delta. (b) Desirability ramp for POE.

Table 9.3 Best Setup for the Lathe

Factor	Setting
A: Speed (fpm)	544
B: Feed (ipr)	0.022
C: Depth (in.)	0.067
y: Delta	0.00
POE (delta)	0.1

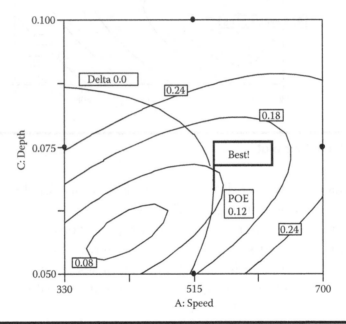

Figure 9.8 Best setup for the lathing product on spec consistently (feed maximized).

The manufacturing manager will like this outcome because it puts the feed rate at its maximum with product right on spec as consistently as possible—given the natural variations in inputs and the process as a whole. Figure 9.8 overlays contours for the delta and its POE with the best setup flagged.

It doesn't get any better than this!

PRACTICE PROBLEMS

9.1 From the website for the program associated with this book, open the software tutorial titled "*Multifactor RSM*".pdf (* signifies other

Table 9.4 Standard Deviations in Input Factors

Factor	Test Limits	Range (Δ)	Standard Deviation (σ)	Relative Variation (σ/Δ) (%)
A: Time (minutes)	40–50	10	0.5	5
B: Temperature (deg C)	80–90	10	1	10
C: Catalyst (%)	2–3	1	0.05	5

characters in the file name) and page forward to Part 3. This is a direct follow-up to the tutorial that we directed you to in Problem 6.1, so, be sure to do that one (Part 2) first. This tutorial, which you should do now, demonstrates the use of the supplied software for doing POE. It requires the details on factor variation provided in Table 9.4.

This tutorial exercises a number of features in the software; so, if you plan to make use of this powerful tool, do not bypass it. However, even if you end up using a different software, you will still benefit by poring over the details on how to generate an operating setup that not only puts the process in its sweet spot relative to all specifications, but also at a point that will be most robust to variations transmitted from the input factors.

9.2 Consider the following standard deviations in the input factors for the DOE golfer discussed in Problem 8.2:

A. Length, plus or minus 1 inch

B. Angle, plus or minus 1 degree

C. Weight, plus or minus 0.1 (assume that washers vary in weight)

There are many setups that achieve the goal of 72 inches, but will some of them result in less POE transmitted from these variations?

Appendix 9A: Details on How to Calculate POE

POE applied to RSM is done most efficiently with the aid of matrix algebra. The first step is a big one: do your DOE and fit an appropriate model to the response. Use the mean square of residuals from the ANOVA to estimate the underlying error variance (σ_e^2), and estimate the factor variances (σ_{kk}^2). Then proceed with POE as follows:

1. Set up a matrix of factor variances:

$$\sigma_1, \sigma_2, \ldots, \sigma_k \implies \Sigma = \begin{pmatrix} \sigma_{11}^2 & 0 & 0 \\ 0 & \sigma_{22}^2 & 0 \\ 0 & 0 & \sigma_{kk}^2 \end{pmatrix}$$

2. Calculate variance

$$\mathrm{Var}(y) = \nabla^{\mathrm{T}} \Sigma \nabla + \sigma_e^2$$

where

$$\nabla = \begin{pmatrix} \dfrac{\partial f}{\partial x_1} \\[2mm] \dfrac{\partial f}{\partial x_2} \\[2mm] \dfrac{\partial f}{\partial x_k} \end{pmatrix}$$

$$\mathrm{Var}(y) = \begin{pmatrix} \dfrac{\partial f}{\partial x_1} & \dfrac{\partial f}{\partial x_2} & \dfrac{\partial f}{\partial x_k} \end{pmatrix} \begin{pmatrix} \sigma_{11}^2 & 0 & 0 \\ 0 & \sigma_{22}^2 & 0 \\ 0 & 0 & \sigma_{kk}^2 \end{pmatrix} \begin{pmatrix} \dfrac{\partial f}{\partial x_1} \\[2mm] \dfrac{\partial f}{\partial x_2} \\[2mm] \dfrac{\partial f}{\partial x_k} \end{pmatrix} + \sigma_e^2$$

$$\mathrm{Var}(y) = \left(\dfrac{\partial f}{\partial x_1} \right)^2 \sigma_{11}^2 + \cdots + \left(\dfrac{\partial f}{\partial x_k} \right)^2 \sigma_{kk}^2 + \sigma_e^2$$

3. Complete the calculation for POE by taking the square root of the variance in response y:

$$\mathrm{POE} = \sqrt{\mathrm{Var}(y)}$$

Some things to note about our calculations:

- All units of measure for factors are actual, and not coded.
- The POE is in the same units as the response measurements.
- They assume no covariance in the factors (all the off-diagonal elements in the matrix are zero); in other words, each factor's noise is independent of all the other factors' noise.
- They end at the second moment whereas others go on the third moment (Taylor, 1996), which produces a slightly higher result, but makes no difference for finding the flats—the ultimate objective for the application of POE to RSMs.

AN IMPORTANT CAVEAT ON POE

Up until now, we've always assumed that variation in the input factors (the x's) will be constant regardless of their absolute level. For some factors, it may be more natural for the error in x to be a constant percent. In this case, the advantage of repositioning the factor in respect to the POE may be negated as shown in Figure 9A.1

Determining the variations in factors may prove to be the key for developing a valid prediction on POE. Do not assume that these will be constant across the operable ranges.

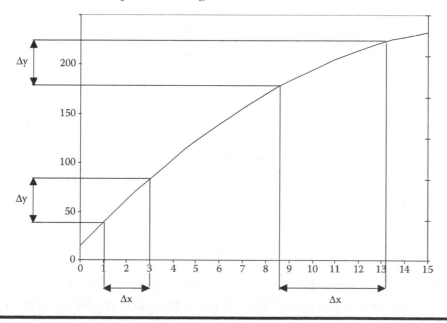

Figure 9A.1   Constant percent error for the input factor x.

POE for Transformed Responses

Occasionally, for statistical purposes, it becomes necessary to transform responses via mathematical functions such as the logarithm, square root, or the like (refer to *DOE Simplified*, Chapter 4: "Dealing with Non-Normality via Response Transformations"). Because RSM designs are geared to fit nonlinear surfaces, transformations are less likely to be beneficial than for screening studies via two-level factorials. That's good because, as you will see, transformations create complications for the calculation of POE.

The transformed response, labeled y with a prime, is first fit to a polynomial function of the input factors (x's) in the usual manner for RSMs. The POE in a transformed scale is then represented by

$$\text{Var}(y') = \sigma_{y'}^2 = \sum_{i=1}^{k} \left(\frac{\partial y'}{\partial x_i} \right)^2 \sigma_{X_i}^2 + \sigma_{resid}^2$$

To get the response back into the original scale (y), the inverse of the transformation must be applied (e.g., an antilog on the logarithm). That's simple compared to what's needed to convert POE into the original scale—multiplication by the partial derivative of y with respect to y' ($\delta y / \delta y'$) and application of the chain rule (from calculus):

$$\frac{\partial y}{\partial x} = \frac{\partial y}{\partial y'} \cdot \frac{\partial y'}{\partial x}$$

This yields

$$\text{Var}(y) = \sigma_y^2 = \sum_{i=1}^{k} \left(\frac{\partial y}{\partial x_i} \right)^2 \sigma_{X_i}^2 + \left(\frac{\partial y}{\partial y'} \right)^2 \sigma_{resid}^2$$

Finally, to compute the POE:

$$\text{POE} = \sqrt{\text{Var}(y)}$$

For example, assume that you do a simple one-factor RSM and collect the actual data as shown in Figure 9A.2. A good fit is accomplished with a quadratic model on the square root of the response.

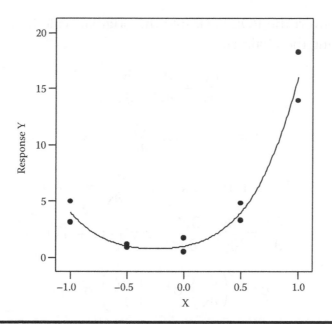

Figure 9A.2   Data from one-factor RSM fitted to the square root of the response.

The model in transformed units ($y' = \sqrt{y}$) is

$$y' = 1 + 1x + 2x^2$$

with a residual standard deviation (σ_{resid}) of 0.27.

The experimenter measures a standard deviation for the factor (σ_x) of 2. First, we calculate POE in the transformed scale

$$\sigma_{y'}^2 = \sum_{i=1}^{k}\left(\frac{\partial y'}{\partial x}\right)^2 \sigma_x^2 + \sigma_{resid}^2$$

$$\frac{\partial y'}{\partial x} = \frac{\partial\left(1 + 1x + 2x^2\right)}{\partial x} = 1 + 4x$$

$$\sigma_{y'}^2 = (1 + 4x)^2(2)^2 + (0.27)^2$$

$$POE_{y'} = \sqrt{(1 + 4x)^2(2)^2 + (0.27)^2}$$

Next, we convert the POE back into the original scale by multiplying $\partial y/\partial y'$ and using the chain rule:

$$\frac{\partial y}{\partial x} = \frac{\partial y}{\partial y'} \cdot \frac{\partial y'}{\partial x}$$

$$y = (y')^2 \quad \therefore \quad \frac{\partial y}{\partial y'} = 2(y')$$

$$y' = 1 + 1x + 2x^2 \quad \therefore \quad \frac{\partial y'}{\partial x} = 1 + 4x$$

$$\frac{\partial y}{\partial x} = \frac{\partial y}{\partial y'} \cdot \frac{\partial y'}{\partial x} = 2(y') \cdot (1 + 4x)$$

$$\sigma_y^2 = \left(\frac{\partial y}{\partial x}\right)^2 \sigma_x^2 + \left(\frac{\partial y}{\partial y'}\right)^2 \sigma_{\text{resid}}^2$$

$$\sigma_y^2 = (2(y')(1 + 4x))^2 \sigma_x^2 + (2(y'))^2 \sigma_{\text{resid}}^2$$

$$\text{POE} = \sqrt{\left(2\left(1 + 1x + 2x^2\right)(1 + 4x)\right)^2 (2)^2 + \left(2\left(1 + 1x + 2x^2\right)\right)^2 (0.27)^2}$$

Wasn't that fun? The plot of POE is shown in Figure 9A.3. When all is said and done, does this picture help you to locate the flats in the actual response?

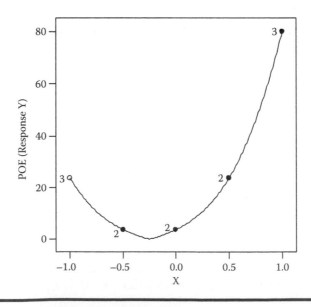

Figure 9A.3   POE from a transformed response.

Yes, it does! Notice that the minimum in POE occurs where the response is the most stable (as seen in Figure 9A.2). This would be a good place to set the input factor x where its variation will be transmitted at the least—a goal consistent with the philosophy of Six Sigma.

"POE" NOT POE: FINAL THOUGHTS FOR THE MORE LITERARY READERS

POE is a very esoteric tool. For those who do not deal with numbers, it perhaps might be best described as "mathemagical." Here's a quote from the famous author of the same name (Poe). Did he disrespect statistics and calculus? You be the judge:

> There are few persons, even among the calmest thinkers, who have not occasionally been startled into a vague yet thrilling half-credence in the supernatural, by coincidences of so seemingly marvelous a character that, as mere coincidences, the intellect has been unable to receive them ... such sentiments are ... thoroughly stifled by the doctrine of chance ... technically termed the Calculus of Probabilities.

> ***The Mystery of Marie Roget***
> A sequel to Poe's more renowned novel
> *The Murders on the Rue Morgue*

Chapter 10

Other Applications for RSM

I'm very well acquainted, too, with matters mathematical, I under-
stand equations, both the simple and quadratical.

Gilbert and Sullivan
A verse from "I Am the Very Model of a Modern
Major General" from *The Pirates of Penzance*

If you've made it this far, you are indeed well acquainted with matters that
are mathematical, particularly those dealing with quadratic equations for
response modeling. All this is easy with numeric (quantitative, continuous)
factors, but what if some of them are categorical (qualitative), such as the
type of material or choice of the supplier? That's one of the issues we will
address in this chapter.

We also discuss the application of RSM to computer simulations, a neces-
sity for modern major generals and the suppliers of their high-tech systems.
For example, it would be prohibitively expensive to build prototype jet
engines for state-of-the-art military aircraft; so, instead, the engineers apply
computer-based techniques such as finite-element analysis to evaluate alternative designs. One compressor may require more than 100,000 elements,
which consume considerable time on high-cost computers (Myers et al.,
2016, p. 564). RSM quickly and inexpensively ferret out transfer functions
(polynomial models) that simplify the search for optimal factor settings.

Adding Categoric Factors to RSM Designs

The developers of RSM, Box and Wilson, came out of the chemical industry, as did the authors of this book. Numeric factors abound for this application of RSM, and they are very easily adjusted. For example, one of the authors (Mark) began his chemical engineering career on a high-pressure hydrogenation unit. The key factors were residence time in the continuous reactor, temperature, and pressure. All these variables are numeric; that is, they can be adjusted to any level over a continuous operating range. However, other factors in this hydrogenation process, mainly related to the catalyst, were categoric in nature—either one level or the other could be selected, but nothing in-between. For example, the reactor might be charged with palladium or rhodium catalyst coated on carbon or zeolite substrate, thus creating four discrete combinations. To complicate things further, similarly designed catalysts could be purchased from a variety of suppliers, but only one would win out in the end.

**THE RISKY BUSINESS OF RSM ON
HIGH-PRESSURE PROCESSING**

The R&D center where the authors first worked thought it best to locate high-pressure reactors outdoors on what they ominously called the "slab." For some reason, they had problems getting senior scientists to volunteer for work on these extremely volatile processes so that they were relegated to the temporary interns known as summer engineers. Miraculously, despite the inexperience and overenthusiasm of these budding technical professionals and the tortuous combinations of factors they subjected these high-pressure units to in the name of RSM, no major disasters ensued. The only rumbles heard on the back-lot location of the slab were the freight trains that passed by frequently on the tracks just on the other side of the fence.

Safety must always be a concern when working around machinery—particularly so when processing flammable materials at high temperature and pressure. Always apply a high degree of caution when venturing into previously unexplored operating regions. Be ready to pull the plug at the earliest indication that things are going out of control!

As you can imagine, the introduction of categoric factors to RSM increases the complexity exponentially. Let's step back a bit and make this as simple as possible by considering what happens if you add only one categorical factor and restrict it to two levels ("treatments" in the terminology of classical DOE). Rather than discuss this in theory, we will make use of a case study to show how to deal with the situation.

A catheter produced for use in a medical device must meet exacting specifications, 2.45 plus or minus 0.05 millimeters, on its outside diameter (OD) or it will not properly mate with other parts. The catheter is made via an extrusion process that creates varying OD depending on three critical factors:

A. Extruder feed rate: 900–1100 grams per hour
B. Zone-one temperature: 165–195 degrees Celsius
C. Raw material vendor: Acme or Emca

Obviously, the last factor is categoric. This could be simply handled by performing two separate RSM designs on A versus B—one with Acme and the other using Emca material. Then, go head to head with the optimal outcomes for each and let the best vendor win. This approach boils down to making two response surface maps and comparing them side by side.

Unfortunately, although this is an easy way out of confronting categoric factors, it wastes an opportunity to learn how they interact with numerical factors, and more importantly, it stymies statistical analysis of their main effects. Somehow, we need to integrate these discrete variables into the polynomial models and include them and their interactions in the ANOVA, so that they can be tested for significance. This can be done very easily by coding the treatments numerically. For Acme versus Emca, pick two numbers, preferably –1 versus +1. This coding is standard for two-level factorial designs; so, it's natural for RSM, which is often built up from these simpler experiments.

The RSM design for the catheter optimization is laid out in Table 10.1 with both actual and coded levels shown.

Notice that the numeric factors are restricted to three values within the ranges specified at the outset of this case study. This can be accomplished via a FCD as depicted in Figure 10.1.

This looks quite different from the normal FCD on three factors, because one of them can only be set at two categoric levels—represented from left to right. In essence, this design breaks down into two FCDs, one for Acme and the other for Emca. (For what it's worth, a two-factor FCD is equivalent to a full three-level factorial design.)

Table 10.1 Data for RSM on the OD of a Catheter

Std	A: Feed Rate grams/hour	Coded	B: Temperature degree Celsius	Coded	C: Vendor Actual	Coded	OD (millimeters)
1	900	−1	165	−1	Acme	{−1}	2.07
2	1100	1	165	−1	Acme	{−1}	2.12
3	900	−1	195	1	Acme	{−1}	2.28
4	1100	1	195	1	Acme	{−1}	2.48
5	900	−1	180	0	Acme	{−1}	2.22
6	1100	1	180	0	Acme	{−1}	2.35
7	1000	0	165	−1	Acme	{−1}	2.13
8	1000	0	195	1	Acme	{−1}	2.39
9	1000	0	180	0	Acme	{−1}	2.35
10	1000	0	180	0	Acme	{−1}	2.31
11	1000	0	180	0	Acme	{−1}	2.34
12	1000	0	180	0	Acme	{−1}	2.37
13	1000	0	180	0	Acme	{−1}	2.36
14	900	−1	165	−1	Emca	{+1}	2.32
15	1100	1	165	−1	Emca	{+1}	2.27
16	900	−1	195	1	Emca	{+1}	2.38
17	1100	1	195	1	Emca	{+1}	2.55
18	900	−1	180	0	Emca	{+1}	2.41
19	1100	1	180	0	Emca	{+1}	2.48
20	1000	0	165	−1	Emca	{+1}	2.39
21	1000	0	195	1	Emca	{+1}	2.44
22	1000	0	180	0	Emca	{+1}	2.49
23	1000	0	180	0	Emca	{+1}	2.44
24	1000	0	180	0	Emca	{+1}	2.49
25	1000	0	180	0	Emca	{+1}	2.47
26	1000	0	180	0	Emca	{+1}	2.47

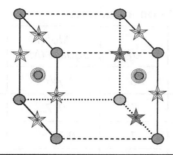

Figure 10.1 FCD for a study with two numeric factors and one categoric factor.

RSM ACHIEVES THE ACME OF PROCESS PERFORMANCE

The designation "Acme" as a generic name for a manufacturer is inextricably linked to Wile E. Coyote and his never-ending quest to get the best of Road Runner. Invariably, WEC finds himself neatly sidestepped by RR at the very pinnacle, or acme, of the desert outcropping. The Acme technology that propelled this determined canine past his prey always fails at the highest point. What a pleasure it is to see the not so Wile E. Coyote plummet past the innocent Road Runner! Oh, oh, it seems that we are sliding down the slippery slope of pure statistics to pure sadistics.

Table 10.2 shows the ANOVA for the chosen model of OD—a quadratic polynomial. Notice that it lacks the term C^2, which does not exist in this case due to this factor (supplier) being categorical. However, by integrating the choice of a supplier into the design and analysis, we now see that it not only creates a significant main effect, but it also interacts with the feed rate (A) and temperature (B). In other words, how the process should be set up in terms of A and B will depend on which vendor the purchase agent chooses at any given time (presuming both will work, which we must investigate).

Residual diagnostics for this model pass all tests:

■ Normal plot lined up as expected
■ No pattern versus predicted response or by run order
■ No trends with run order
■ Outliers are not evident
■ Box–Cox plot displays no significant improvement for any of the power transformations

Table 10.2   ANOVA for Data on a Catheter

Source	Sum of Squares (SS)	df	Mean Square (MS)	F Value	p-Value Prob > F
Model	0.36	8	0.045	79.72	<0.0001
A	0.027	1	0.027	42.2	<0.0001
B	0.12	1	0.12	193.	<0.0001
C	0.13	1	0.13	201.	<0.0001
AB	0.017	1	0.017	26.7	<0.0001
AC	0.0030	1	0.0030	4.69	0.0449
BC	0.016	1	0.016	25.1	0.0001
A^2	0.0074	1	0.0074	11.6	0.0034
B^2	0.023	1	0.023	35.4	<0.0001
Residual	0.011	17	0.00064		
Lack of fit	0.0071	9	0.00079	1.66	0.242
Pure error	0.0038	8	0.00047		
Cor Total	0.37	25			

Figure 10.2a and b shows contour plots for Acme versus Emca with the target shown via a dotted line, and outer specifications in bold.

It now becomes obvious that catheters made from either vendor's raw material can be produced to the targeted OD—provided the process gets set up correctly for feed rate versus temperature. Emca is clearly preferred since it works over a broad swath of process conditions—it will be most robust to variations in inputs. Is there a setup where both vendors work? This would be very desirable to achieve stable operations while allowing the purchase agent to encourage price competitiveness. The answer becomes apparent in Figure 10.3, an overlay plot that displays the in-specification window or sweet spot. This graphical optimization tool reveals that yes, it may be possible to run the catheter process at one setup of feed rate (A) and temperature (B) and let the purchase agent switch vendors at any time. However, although it may be in spec, the part will probably be consistently off the target of 2.45 millimeters OD—either high or low depending on which material (C) happens to be fed in at any given time.

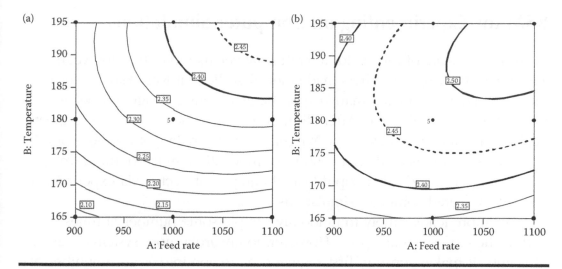

Figure 10.2 **(a) Contour plot for Acme. (b) Contour plot for Emca.**

In this case, it pays to reset the process whenever the vendor is switched, thus getting the part back on target.

As you can see from this example, categorical factors can be accommodated in RSM, but they create a lot more work and introduce complications for analysis. Therefore, it may be best to screen out categoric factors in the earlier stages of experimentation and hold them fixed for the final optimization via RSM.

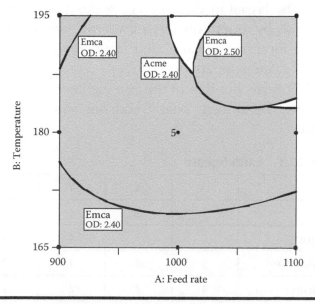

Figure 10.3 **Graphical optimization plot for both vendors.**

RSM for Experimenting on Computer Simulations

Consider the state-of-the-art for aircraft now versus more than 100 years ago when the Wright brothers made their first flight at Kitty Hawk, North Carolina. They conducted many experiments on wing design, propeller configuration, and the like. After all these preliminaries, it still took a great deal of trial and error before the Wrights' aircraft finally got off the ground. Nowadays—after more than a century of innovation, much of the development for aircraft and other sophisticated equipment occurs via experiments on high-powered computer simulations.

One approach is simply to randomly sample some number of times within the experimental space. However, to ensure a more systematic array of points, it makes sense to first segment the region into a given number of rows and columns. Then, sampling can be done in such a way that, in each row and each column, one point appears—no more, no less. This result is called a Latin hypercube design or LHD (McKay et al., 1979).

HISTORY OF LATIN SQUARE

A Latin square is simply a grid, in which each number appears only once in each row and column. Table 10.3 shows a Latin square of order 4.

These arrays make good templates for the DOE with multiple blocking. For example, suppose you wanted to check test four brands of tires: W, X, Y, and Z. The layout in Table 10.3 provides a sensible plan for rotating the tires month by month on varying wheels identified by numbers 1 through 4 by position on the car: 1. left front, 2. right front, 3. left back, and 4. right back (Peterson, 2000).

Why "Latin"? It turns out that in the late eighteenth century, the mathematician Euler (pronounced "oiler") laid out squares of this sort out using Latin letters.

Table 10.3   Latin Square

	W	*X*	*Y*	*Z*
July	1	2	3	4
August	2	1	4	3
September	3	4	1	2
October	4	3	2	1

> For since the fabric of the universe is most perfect and the work of a most wise Creator, nothing at all takes place in the universe in which some rule of maximum or minimum does not appear.
>
> **Leonhard Euler**

Figure 10.4 shows an LHD cited by Myers and Montgomery (2009, p. 484) for two factors, each ranging from 0 to 16. The levels in each column of the design matrix are randomly arranged to construct the design.

As you can see, this semirandom scattering of points leaves much to be desired for filling all the space defined by the experimental factors. We suggest better alternatives in Appendix 10A.

When the objective is to estimate a polynomial transfer function, we shift from the LHD to more traditional designs for RSM, such as the CCD. For example, NASA engineers developed a computer simulation of a tetrahedral truss for a scientific space platform (Unal et al., 1997). They studied nine factors in only 83 runs via a CCD with a fractional factorial core and rotatable star points (alpha 2.83).

Other aerospace engineers made use of a FCD to improve the wing design on a lightweight fighter jet (Zink et al., 1999) (see Table 10.4). They wanted to assess a new Active Aeroelastic Wing (AAW) technology

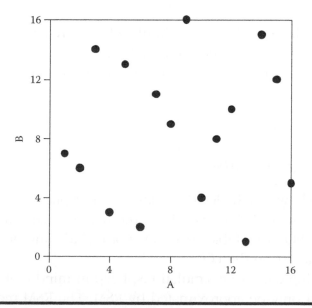

Figure 10.4   Latin hypercube on two factors.

Table 10.4 FCD on the Wing Design for a Jet Fighter

Run	A: Aspect	B: Taper	C: Thickness	Wing Weight (pounds)
1	3	0.2	0.03	334.6
2	3	0.2	0.06	126
3	3	0.4	0.03	407.3
4	3	0.4	0.06	161.2
5	5	0.2	0.03	833.7
6	5	0.2	0.06	392.1
7	5	0.4	0.03	1070.2
8	5	0.4	0.06	326.5
9	3	0.3	0.045	226
10	5	0.3	0.045	460.9
11	4	0.2	0.045	286.9
12	4	0.4	0.045	408.3
13	4	0.3	0.03	608.5
14	4	0.3	0.06	236.7
15	4	0.3	0.045	380.5

(see Figure 10.5) that could only be simulated via physics-based finite-element analysis on high-powered computers.

The simulator generated an estimate of wing weight, which the engineers hoped to minimize as a function of three key factors:

A. Aspect ratio: 3–5
B. Taper ratio: 0.2–0.4
C. Thickness ratio: 0.03–0.06

This design, shown in Table 10.4, contains no replicates because they would generate identical responses from the deterministic simulation. Therefore, the ANOVA in Table 10.5 does not include any pure error, nor does it provide a test on the LOF.

When analyzing data from simulations, keep in mind that the true computer model will only be approximated by RSM. The RSM metamodel will not only fall short in the form of a model, but also in the number of

Figure 10.5   NASA's AAW F/A-18A research aircraft.

factors. Therefore, predictions will exhibit a systematic error, or bias. This is what will be measured in the residual—rather than the normal variations observed from experiments on physical processes. Despite these circumstances, much of the standard statistical analyses remain relevant, including measures of model fit such as PRESS and R^2_{Pred}. However, the p values will not be accurate estimates of risks associated with the overall model or any of its specific terms.

Table 10.5   ANOVA on the Wing Design

Source	Sum of Squares	df	Mean Square	F Value	p-Value Prob > F
Model	867,623	6	144,603	47.4	<0.0001
A	334,268	1	334,268	109.5	<0.0001
B	16,016	1	16,016	5.2	0.0512
C	404,733	1	404,733	132.6	<0.0001
AC	66,722	1	66,722	21.9	0.0016
BC	14,416	1	14,416	4.7	0.0615
C^2	31,467	1	31,467	10.3	0.0124
Residual	24,416	8	3052		
Cor Total	892,039	14			

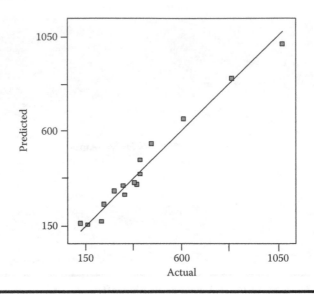

Figure 10.6 Actual response from a simulator versus predicted response.

Table 10.6 Statistics for Models on Wing Simulation, with and without Log Transformation

Statistic	No Transform	Log Transform
Model	Quadratic	Linear
F-value	20.1	110.5
$R^2_{Predicted}$	0.8627	0.9362

The fit of predicted versus actual data (from the simulation) looks very good as you can see in Figure 10.6.

However, the Box–Cox plot (see sidebar "The Box–Cox Plot") shows that residuals can be significantly reduced via a log transformation, which is not surprising considering the broad range of response—from 150 to 1050. Furthermore, as shown in Table 10.6, the resulting model, with the response in log scale, is more parsimonious (what a great word for saying "simpler"!). Even though it contains only linear terms (A, B, and C), the log model fits the actual data more precisely.

THE BOX–COX PLOT

We already detailed the Box–Cox plot in Appendix 5A, but we thought it would be worth reviewing. As you may recall, this graphical method, named after its originators, shows how dimensionless residuals change

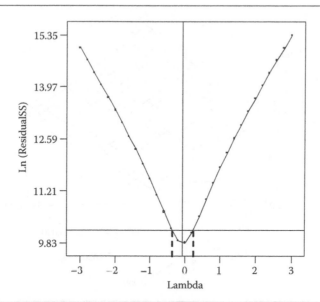

Figure 10.7   Box–Cox plot for a linear model for wing performance.

as a function of varying powers of response transformation. Figure 10.7 shows the results for fitting the wing simulation.

The y-axis displays the residuals after transforming the response by a range of powers from −3 (inverse cubed) to +3 (cubed). The powers, labeled as "Lambda" (Greek letter λ), are displayed on the x-axis of the Box–Cox plot. The option of not doing any transformation is represented by the value 1. This specifies that all responses (y) be taken to the power lambda of 1, but y^1 equals y, so that this makes no difference.

In this case, the plot shows the power of 1 being outside of the 95% CI for minimal residuals (indicated via the dotted lines). Notice that the minimum residual occurs near the value of 0. This is another special power because it makes no sense to compute y^0, which of course would create all 1's. It actually represents taking the logarithm of the response, which proves to be very effective for modeling the performance of the wing on this particular rocket ship.

Which is more daunting—rocket science or statistics?

A human being is the best computer available to place in a spacecraft...It is also the only one that can be mass produced with unskilled labor.

Werner von Braun

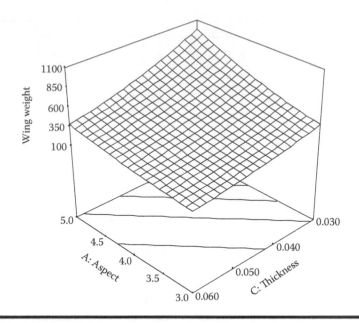

Figure 10.8   3D response surface for wing simulation.

The 3D surface in Figure 10.8 shows how the predicted response varies for the two factors creating the most impact on the wing—aspect (A) and thickness (C).

Pictures like this can far outweigh the untold numbers potentially generated by a computer simulation, oftentimes at a great expense.

PRACTICE PROBLEM

10.1 From the website for the program associated with this book, follow the "Optimal Categoric" link to the software guide titled "Multifactor RSM Tutorial (Adding Categorical Factors)." It parallels a study done by Procter and Gamble (P&G) engineers on a new sealing process (Brenneman and Myers, 2003). They were concerned about how the maximum peel strength would be affected by changing suppliers of the packaging material. They set up an RSM design to vary several key factors on the sealing machine, including the supplier:

A. Temperature: 193–230 degrees Celsius
B. Pressure: 2.2–3.2 bar
C. Speed: 32–50 cpm (cycles per minute)
D. Supplier: S1, S2, and S3

Owing to limitations on time and other resources, a maximum of 37 runs could be performed. Therefore, simply doing a standard CCD or BBD for each of the three suppliers would not do—these design choices produce far too many runs (60 and 51, respectively). Instead, the P&G engineers made use of an optimal design.

Using the software provided with this book, we laid out the experiment, shown in Table 10.7. The data come from a simulation that's loosely based on the predictive model reported in the cited article. (Some liberties were taken to make the outcome more interesting.) Assume that the maximum peel strength will ideally hit a target of 4.5 pound force (lbf). However, it must exceed 3 lbf to prevent leaking and not go above 6 lbf because the package becomes too difficult to open.

The tutorial explains how this design was constructed via optimal criteria with added points for LOF testing and pure-error estimation. Notice that we made use of nearly all the budgeted runs. Also, aided by a feature in the software that forces balance, we made sure that each supplier got an equal number of design runs.

Use the provided software, with guidance from the tutorial, to analyze the responses and then find desirable solutions for the following objectives—listed according to the relative acceptability for all concerned:

A. Can the process be adjusted to hit the maximum peel strength target of 4.5 lbf for any one of the suppliers?

B. The purchase agent intended to break down the supply in this manner: 50% to S1, 25% to S2, and 25% to S3. Can you find a process setup that will work for all suppliers? If not, a two-supplier option might be satisfactory, provided S1 is one of them. In other words, assuming it will not be robust for all three suppliers, can the process be set up in such a way that either S1–S2 or S1–S3 meets specifications?

C. If it is not possible to achieve a common setup for even two of the three suppliers, perhaps, the upper limit of 6 could be raised on the maximum peel strength specification. (Assume that customers are willing to use a pair of scissors or, better yet, the package designers can add a notch in the plastic for easier opening.) Will this open up a window of operability for multiple suppliers at one set of process conditions?

Table 10.7 Data from a Package-Sealing Experiment

Setup	A: Temperature (degree Celsius)	B: Pressure (Bar)	C: Speed (cpm)	D: Supplier	Peel Strength (lbf)
1	193	2.2	32	S1	4.6
2	230	2.2	32	S1	10.0
3	193	3.2	32	S1	7.2
4	193	3.2	32	S1	6.6
5	230	3.2	32	S1	5.5
6	202.25	2.7	41	S1	8.2
7	230	2.7	41	S1	8.5
8	193	2.2	50	S1	6.7
9	230	2.2	50	S1	11.0
10	230	2.2	50	S1	11.0
11	230	3.2	50	S1	7.8
12	230	3.2	50	S1	7.1
13	230	2.2	32	S2	5.1
14	230	2.2	32	S2	6.7
15	211.5	2.7	32	S2	2.9
16	230	3.2	32	S2	1.7
17	211.5	2.2	41	S2	6.3
18	193	2.7	41	S2	2.0
19	220.75	2.95	41	S2	3.7
20	193	2.2	50	S2	2.4
21	230	2.7	50	S2	4.0
22	230	2.7	50	S2	4.1
23	193	3.2	50	S2	4.8
24	193	3.2	50	S2	5.1
25	193	2.2	32	S3	6.7
26	193	2.2	32	S3	6.7
27	230	3.2	32	S3	8.0
28	230	3.2	32	S3	7.1

(Continued)

Table 10.7 (*Continued*)   Data from a Package-Sealing Experiment

Setup	A: Temperature (degree Celsius)	B: Pressure (Bar)	C: Speed (cpm)	D: Supplier	Peel Strength (lbf)
29	220.75	2.45	36.5	S3	9.2
30	202.25	2.7	36.5	S3	8.1
31	211.5	3.2	41	S3	7.2
32	230	2.2	50	S3	9.5
33	230	2.2	50	S3	9.8
34	211.5	2.7	50	S3	6.2
35	193	3.2	50	S3	5.9
36	230	3.2	50	S3	4.8

Appendix 10A: Alternative RSM Designs for Experiments on Simulations

In Chapter 7, we provided a checklist for gauging the quality of designs for RSM. Many of the line items can be carried forward for application to deterministic computer simulations, but, as shown in the following list, others (struck out in this list) can be ignored—mainly due to the lack of variability in the inputs and outputs:

- Generate information throughout the region of interest
- Ensure that the fitted value, $\hat{y}$, be as close as possible to the true value
- Give a good detectability of the LOF
- Allow experiments to be conducted in blocks
- Allow designs of an increasing order to be built up sequentially
- Require a minimum number of runs
- Unique design points in excess of the number of coefficients in the model chosen by the experimenter
- Remain insensitive to influential values and bias from model misspecification

Santner et al. (2003), who wrote the book on computer experiments, offer a much shorter list. They advise the use of designs that

- Provide information about all portions of the experimental region
- Allow one to fit a variety of models

Surprisingly, these authors suggest that at least one observation in any set of inputs be replicated to guard against unannounced changes in the computer code.

Let's revisit the simple case of two factors for which we showed an application of the LHD. Figure 10.4 displays an LHD with 16 runs—far more than needed to fit a quadratic polynomial, which normally would be adequate for purposes of RSM, but may not be for complex computer simulations.

One alternative to consider, at least for a few factors, is a distance-based design, which can be readily constructed via software that supports RSM. Simply put, the distance criterion selects a given number of points from a candidate set in such a way that they will be spaced as far apart as possible. For example, the following 17-point distance-based design (black circles in Figure 10A.1) results from a candidate set consisting of 29 factorial combinations laid out concentrically by Design-Expert software (open circles for points not chosen).

Beyond two factors, the distance-based design becomes less and less practical due to putting too few points in the interior. For example, a 30-point design on four factors puts only three points in the interior, that is, 10%.

When computer time is very costly, LHD designs become excessive in their number of runs. One school of thought by experienced practitioners of RSM on simulation is that CCDs work well for up to five factors (Unal et al., 1998). This design choice reduces the number of runs considerably, for

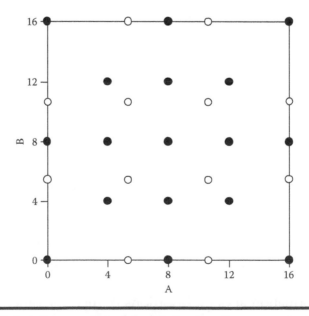

Figure 10A.1   17-Point distance-based design for two factors.

example, to only nine points for the two-factor case (for a CCD with no replicates). However, these same experts (Unal et al., 1998) suggest that for six or more factors, an optimal design geared to the quadratic model is a good choice *if* it is overdetermined by 50% or so. For example, NASA scientists performed 45 runs on a simulator for a rocket-powered single-stage launch vehicle (Unal et al., 1998). Only 29 runs were needed at a minimum for fitting the quadratic model on six factors related to wings and the like on this spacecraft. Therefore, they overdid the design by 55%.

Hitchhiking off this idea, we propose to use an I-optimal criterion to pick the minimal set of candidate points needed for the quadratic polynomial. Then, add half of that many points based on the distance-based criterion. These augmented design points plug the remaining gaps and thus achieve the objective for computer simulations to be space filling.

Another approach that combines optimal with space-filling designs does it in two stages (Johnson et al., 2010):

1. Lay out a number of points via a space-filling design. Fit an adequate polynomial model.
2. Augment this first design using an optimal criterion to the degree of a polynomial as seen fit from step 1.

GOING THE DISTANCE

Here's how the distance-based criterion works:

1. Specify n points such that $p < n < c$, where p is the number of parameters in the chosen model and c is the size of the candidate set
2. Choose an initial design point at a vertex in the experimental space
3. Add the next candidate point whose minimum Euclidean distance[*] from points already in the design is as large as possible
4. Repeat step 2 until all n points are chosen

---

[*] Defined by US National Institute of Standards (NIST) as follows:
The straight line distance between two points. In a plane with p_1 at (x_1, y_1) and p_2 at (x_2, y_2), it is $\sqrt{((x_1 - x_2)^2 + (y_1 - y_2)^2)}$. *In N dimensions, the Euclidean distance between two points p and q is* $\sqrt{(\sum_{i=1}^{N} (p_i - q_i)^2)}$ *where* p_i *(or* q_i*) is the coordinate of p (or q) in dimension i.* (*Source:* www.nist.gov/dads/HTML/euclidndstnc.html.)

This criterion may not provide sufficient points to fit the chosen model. However, the distance algorithm can be modified so that it only picks points that increase the rank of the design matrix until it equals the number of model coefficients. We do not recommend the use of this modified distance approach because optimal selection is far superior for picking model points.

NIST goes on to suggest that you also see "Manhattan distance." This is the distance between two points measured along the axes at right angles, which some people call the "taxicab metric." Unless you are a pigeon, this sounds to be much more practical for city dwellers than Euclidean distance!

Chapter 11

Applying RSM to Mixtures

> If your experiment needs statistics, then you ought to have done a better experiment.
>
> **Ernest Rutherford**
> *Nobel Prize for chemistry 1908*

The attitude of elite chemists toward statistics has not improved much from when Rutherford made this insulting statement. Perhaps, the reason is that the standard methods for the DOE don't work very well on mixtures. For example, let's say you get a new ultra-high-shear blender and start tossing in various fruits to see if you can make a tasty "smoothie" drink. Table 11.1 shows the experimental layout for a fanciful concoction that might be branded "BanApple." Is this a good design?

Aside from the dubious choice of ingredients for this mixture design, it makes no sense when you consider that the taste will be simply a function of the proportions of ingredients. Notice that standard orders 1 and 4 end up being the same in terms of the fractions for each fruit. In other words, all that's been done is a scale-up of the same recipe. Yuk! Who would want to double the dose of a BanApple smoothie? The total amount varies, but will have no effect on responses such as taste, color, or viscosity. Therefore, it makes no sense to do the complete design. When responses depend only on proportions and not the amount of ingredients, factorial designs don't work very well. The same problems occur with RSM, such as the CCD, which uses factorials as its core.

Table 11.1   Bad Factorial Design on Mixture of Fruit for a "Smoothie"

Std Order	A: Apples	B: Bananas	Proportions (A/B)	Fraction (A, B)
1	2	1	2/1	(0.667, 0.333)
2	4	1	4/1	(0.8, 0.2)
3	2	2	1/1	(0.5, 0.5)
4	4	2	2/1	(0.667, 0.333)

A MIXTURE BY ANY OTHER NAME BEHAVES THE SAME

Mixtures don't necessarily refer only to physical substances. For example, varying effects can be produced by changing the mix of words in a paragraph. However, writers such as William Safire of the *New York Times*, who pride themselves on their command of the English language, find the word *mixture* lifeless. They prefer the word *farrago* (pronounced fuh-RAY-go or fuh-RAH-go). Farrago, which comes from a Latin word referring to a mixed grain used for making mush (yum!), is synonymous with *gallimaufry* derived from Old French. However, farrago is as far as we will go into the realm of the esoteric vocabulary for the erudite experimenter.

...what we have been treated to is a farrago of half-truths, assertions and over-the-top spin.

Peter Kilfoyle
Former British Labor Minister

William Safire column "On Language" in The New York Times Magazine, April 27, 2003, p. 22

Accounting for Proportional Effects by Creating Ratios of Ingredients

To many formulators, the ratios of components mean more than the proportions. For example, in the manufacturing of glass, the ratio of silica to alkali has long been considered to be a key factor for product performance (Sullivan and Taylor, 1919). Similarly, the quality of bread dough greatly depends on the flour-to-water ratio (Veal and Mackey, 2000). By converting multiple components into various ratios, experiments involving formulation can then be run using factorial, central composite, or other process designs. In other words, you can mix your cake and bake it too! However, as we will detail, there are downsides to this approach for the mixture design:

■ It takes some mathematical finagling to develop ratios that will not violate the constraint that ingredients add up to a fixed total, for example, 100%

■ This mathematical finagling must be undone to back calculate from the ratios laid out in the design matrix to actual levels of ingredients (more math ☹)

■ The layout of design points in ratio space often translates into a poorly spaced set of formulations.

GETTING A GRIP ON A SLIPPERY FORMULA

In 1953, while trying to develop a missile-part degreaser, rocket chemical technicians made 39 formulations, none of which worked. But number 40 worked like a charm so that they named their product WD-40 as an abbreviation for "water displacement," perfected on the "40th" try. Noticing that employees started sneaking it home for personal use, the company started selling it to consumers. Over the years, WD-40 has been put to many uses, but none of them have been so unusual; at least, that's been publicized, as the time that firemen needed it to extract a nude burglar from an air vent. The methods used to develop WD-40 and experiments on how it might be put to the best use remain largely unknown, probably for the better.

www.WD40.com

First of all, let's address the issue of forming proper ratios. This will not be a problem if you wish to experiment on only two ingredients. For example, it's not a big deal to make a variety of BanApple smoothies at varying ratios of bananas to apples. An erstwhile entrepreneur could add processing factors, such as blender speed, and create a response surface design aimed at optimizing consumer response. Who knows, maybe, people prefer smoothies that are somewhat lumpy! However, as soon as you add a third ingredient (how about something with some tartness, such as clementines!), the setup of proper ratios starts getting complicated. We must follow certain rules for doing this:

1. The number of ratios (n_r) is equal to $q - 1$, where q represents the number of ingredients (or components in the jargon of the mixture design)
2. Each ratio in the set must contain at least one of the components used in at least one of the other ratios belonging to the set

This latter rule allows the total constraint (often 100%) to be maintained (take our word on this!). Here are several feasible ratios (R_i) for three ingredients (A, B, and C):

- $R_1 = A/C$, $R_2 = B/C$
- $R_1 = A/B$, $R_2 = B/C$
- $R_1 = A/(B + C)$, $R_2 = B/C$

Which of these makes most sense entirely depends on the application and what's been already established as a common practice. This is likely to depend on the chemistry of the formulation. If you really do not care in one way or another, consider labeling the components in a descending order of concentration and then applying the first set of ratios (R_1:A/C and R_2:B/C). This protocol produces ratios greater than one, which you may find more convenient to apply for experimental purposes. For example, in our BanApple smoothie with clementines, assume that we want more apple (A) than banana (B), with clementines (C) being the least of all three ingredients. Then, for purposes of experimentation, two ratios, R_1 (A/C) and R_2 (B/C), could be varied over two levels from low to high. We won't try quantifying this hypothetical DOE because, as you will see in the next example, it requires some arithmetic.

DON'T LIKE OUR BANAPPLE IDEA? HOW ABOUT A FUZZY BANANA NAVEL!

Try this recipe for a soothing sipper:

- Two medium, ripe DOLE bananas, quartered
- One pint DOLE orange sorbet or two cups of orange sherbet, slightly softened
- One cup DOLE mandarin tangerine juice

Combine bananas, sorbet, and juice in a blender or food processor. Blend until it is thick and smooth. It takes only 5 minutes to prepare and serves four with delicious drinks containing 220 calories, 2 grams of fat (1 gram saturated), 5 milligrams of cholesterol, 38 milligrams of sodium, 1 gram of carbohydrate, and 2 grams of protein.

Do you care to venture a guess as to the source of this recipe? (Hint: It's a company that was founded in Hawaii in 1851. They are now the world's largest producer and marketer of high-quality fresh fruit.)

Now, we are ready to illustrate the use of ratios in RSM on an entirely different (not a beverage!) example—blending gasoline (Cornell, 2002, p. 307). A refinery produces three components ($q = 3$) for automotive fuel. Their ratios of interest are C/A and C/B, respectively. These two ratios satisfy the two rules for feasibility, that is,

1. Number of ratios $n_r = q-1 = 3-1 = 2$
2. One common component in both ratios: C

The blending operation currently operates at ratios:

■ $R_1 = C/A = 1$
■ $R_2 = C/B = 2$

This translates to an actual composition in weight fraction for A, B, and C of 0.4, 0.2, and 0.4. The fuel formulators want to vary each ingredient within the following individual constraints:

A. 0.25–0.60
B. 0.1–0.4
C. 0.2–0.6

The weight fractions for three components must always sum to a total of 1. We mustn't forget this!

For purposes of optimization, the petroleum chemist responsible for the gasoline product development creates a full three-level factorial design (3^2) based on the following ratios:

■ R_1 (C/A) = 0.5, 1.0, and 1.5, versus
■ R_2 (C/B) = 1, 2, and 3

Notice that these ranges go somewhat below and above the current ratios of ingredients. However, will they conform to the individual component (A, B, and C) constraints? We can answer this vital question by laying out the design on a trilinear graph paper, also known as "ternary" diagrams.

"TURNARY" DIAGRAMS

If you are not a chemist, chemical engineer, material scientist, or the like, you may not be familiar with trilinear graph paper. This is a useful tool for metallurgists for diagramming the various phases for alloys such as those shown in Figure 11.1 for stainless steels produced at 900 degrees Celsius (American Society for Metals, 1992).

The three main components of stainless steel are iron (Fe), chromium (Cr), and nickel (Ni). They can be varied from zero at each side of the triangle to 100% of the total weight at the opposing vertices. The most common type of stainless steel, often used for kitchen flatware, is the one pointed out on the graph: 18-8. Its name reflects the composition of chromium and nickel, respectively. Notice that the point falls 18% of the way from the bottom to the top of the triangle (Cr) and 8% of the distance from the left side to the corner at the right (Ni). Now that these two compositions are fixed, the third (iron) must make up the difference—74% (Fe).

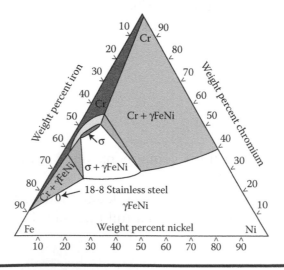

Figure 11.1   Example of a trilinear graph.

Here's a tip that may help you decipher specific compositions pointed out on trilinear graphs: turn the paper so that the ingredient you wish to quantify is oriented with the zero side down and opposing vertex up (hence the pun "turnary" for the proper term—ternary).

Next time you butter your toast, spare a moment to look at the knife (assuming it is 18-8 stainless) and reflect on the wonders of metallurgy and this graphical tool for diagramming alloys.

Figure 11.2 displays the individual constraints and ratios of components for the gasoline-blending example.

Before we discuss the design space, let's first see how to draw in the ratio lines. With simple ratios such as these, it's very easy. For R_1 (C/A), go to the C–A side of the triangle. The ratio of 1 is achieved at the midpoint, or 50/50 level. From there, draw a line to the opposite vertex (B). Along that line, the ratio of one for C/A is preserved. Similarly, you can establish an R_1 (C/A) ratio of 0.5 by choosing one-third (33.3%) of component C versus two-thirds (~66.7%) of A on the same (C/A) side of the trilinear graph. Again, draw a line to the opposite vertex (B). Finally, follow the same process to create a ray for an R_1 of 1.5. Next, we move on to R_2 (C/B). These ratios can be most easily established along the C–B side of the triangle. The ratio of 1 is at the

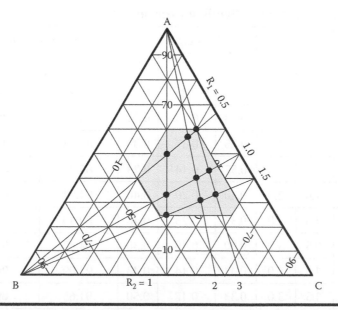

Figure 11.2   Nine gasoline blends for 3^2 RSM design.

50/50 midpoint. That's easy! Again, you can draw a line to the opposite vertex (A) from this point and thus preserve this ratio of 1. By the same procedure, we created rays for R_2 of 2 and 3.

Now comes the fun part: the intersection of the three R_1 rays and the other three rays for R_2 ratios form the design. Notice that the resulting points provide a decent, but not an outstanding, coverage of the feasible mixture space. For example, it does not reach out to the corners, called extreme vertices. Nor does it center the middle points. As we warned you earlier, these less-than satisfactory design layouts are a drawback to the use of ratios, which becomes more pronounced as constraints become more complex. We will offer a better alternative after carrying this example along a bit further. There's still much work to be done for executing and ultimately analyzing this experiment conducted in terms of ratios.

Table 11.2 shows the full three-level design in terms of ratios (R_1 and R_2), the back-calculated recipes (see the sidebar titled "The Tedious Downside of Formulating via Ratios" for details) for the three gasoline components (A, B, and C), and the measured response—the octane number of the resulting fuel. The design is fully replicated in a randomized manner, but for the sake of space, the two results at every unique formulation are tabulated side by side.

Table 11.2   Gasoline-Blending Experiment

Form. ID	R_1 C/A	R_2 C/B	Recipe			Response: Octane	
			A	B	C	First Rep	Second Rep
1	0.5	1.0	0.50	0.25	0.25	84.42	83.16
2	1.0	1.0	0.33	0.33	0.34	86.45	88.66
3	1.5	1.0	0.25	0.37	0.38	84.58	87.64
4	0.5	2.0	0.57	0.14	0.29	85.22	87.96
5	1.0	2.0	0.40	0.20	0.40	94.84	92.07
6	1.5	2.0	0.31	0.23	0.46	89.39	91.71
7	0.5	3.0	0.60	0.10	0.30	84.42	89.12
8	1.0	3.0	0.43	0.14	0.43	91.09	90.73
9	1.5	3.0	0.33	0.17	0.50	91.65	89.51

THE TEDIOUS DOWNSIDE OF FORMULATING
VIA RATIOS: CALCULATING RECIPES

It can be quite a chore to make the necessary translation of ratios, used to design an RSM experiment for formulation, back to the actual composition for use as a recipe sheet by the people doing the actual mixing. In the three-ingredient case for gasoline blending, we have three equations to work with—two for the ratios plus another for the overall constraint on the fixed total (100% or 1 on a scale of zero to one):

1. $R_1 = C/A$
2. $R_2 = C/B$
3. $A + B + C = 1$

Then, with three equations for three unknowns, it's simply (?) a matter of arithmetic* to solve for the three components:

■ $A = R_2/(R_1 + R_1 R_2 + R_2)$
■ $B = R_1/(R_1 + R_1 R_2 + R_2)$
■ $C = R_1 R_2/(R_1 + R_1 R_2 + R_2)$

---

* Suggestion: Make use of readily available software that solves equations like these. Then, whether you calculate by hand or by a computer, check the recipes via a spreadsheet software package to ensure that each formulation adds up to the proper total and produces the specified ratios. This might save much time, trouble, and embarrassment.)

Via least-squares regression, the octane data were fitted to a quadratic polynomial equation to produce this predictive model:

$$\hat{y} = 92.36 + 1.68A + 1.80B - 3.24A^2 - 2.58B^2$$

The 2FI (AB) was insignificant ($p > 0.1$), so it's been removed. All other terms are significant at $p \le 0.05$. The LOF is insignificant ($p > 0.1$) and diagnostics on residuals appear to be normal; so, the model is deemed to be valid for predictive purposes ($R^2_{Pred} = 0.59$). Also, the adequate precision statistic of 10.2 far exceeds the guideline of 4; so, by this measure of signal to noise, the model scores very well. The contour plot (with the optimum

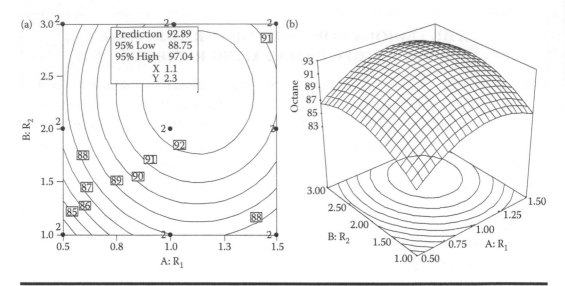

Figure 11.3 **(a) Contour plot for the gasoline case. (b) 3D surface for octane.**

flagged) and 3D response surface are displayed in Figure 11.3a and b, respectively.

The optimum point ($R_1 = 1.13$ and $R_2 = 2.35$) translates (via the equations provided in the sidebar) to a composition of (0.38, 0.19, and 0.43) for (A, B, and C). It produces an octane of 93.

DEALING WITH NONLINEAR BEHAVIOR OF RATIOS

As you've seen in the gasoline-blending case, ratios do not provide a uniform coverage of the mixture space. In Figure 11.4a, you see a graph showing the ratio of A to B versus the level of A. Notice how it blows up as component A goes to a value of 1, because this drives B to 0, causing the ratio to become infinite. This can be counteracted to some extent by transformation with the logarithm (natural or base 10, it will not matter). Figure 11.4b displays a noticeably more linear response of ln(A/B) to A, particularly in the range from 0.2 to 0.8.

Therefore, we suggest that you consider averaging logarithms of the extreme ratios to determine the intermediate ratios. For example, in the gasoline-blending case, the middle values of the two ratios could be transformed as follows:

$$R_1 = e^{((\ln(0.5)+\ln(1.5)/2))} = 0.866 \quad R_2 = e^{((\ln(1.0)+\ln(3.0)/2))} = 1.73$$

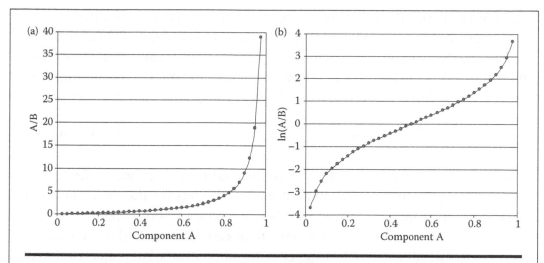

Figure 11.4 **(a) Ratio of two ingredients. (b) Ratio after being logged.**

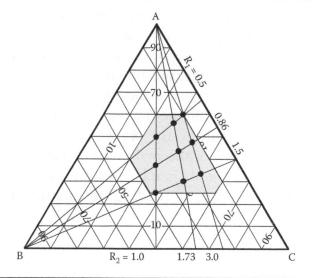

Figure 11.5 **Layout of a gasoline-blending design with new midpoints based on log.**

You can see how this improves the spacing of the middle points in Figure 11.5.

A Different Approach for Optimizing Formulations: Mixture Design

The use of ratios accounts for natural relationships in formulation components, such as the stoichiometry of a reagent to a catalyst in a chemical reaction. However, the predictive models in these terms cannot be interpreted very readily. It would be much handier to see the equation as a function of the original ingredients. This can be done via a polynomial form called Scheffé after the originator (Henri Scheffé, 1958). Here's the predictive model for gasoline octane refitted to the Scheffé polynomial for mixtures:

$$\hat{y} = 75.80A + 56.23B + 73.77C + 39.82AB + 63.88AC + 95.30BC$$

Notice that all three components are detailed in this second-order (nonlinear) equation. Observe that all the coefficients for the interaction terms are positive. This indicates synergism between components—that is, more octane emerges from any two materials than can be expected from a simplistic linear-blending model of the two. In other words, two plus two equals more than four! Formulators are overjoyed when they see synergism like this, the most dramatic of which occurs between components B and C, as evidenced by their model coefficient being the largest of the second-order terms. This is graphically illustrated by the pronounced upward curve in the B–C edge of the 3D response surface graph shown in Figure 11.6b. (The hexagonal region covered is identical to that depicted earlier, except that it's been magnified as far as possible within the boundaries of the trilinear graph.)

Now is a good time to return to the predictive model and observe from inspection of the coefficients for the main effects that material B falls short of the other two. Thus, on the 3D graph in Figure 11.6b, the response dives down toward the B corner of the trilinear mixture space. However,

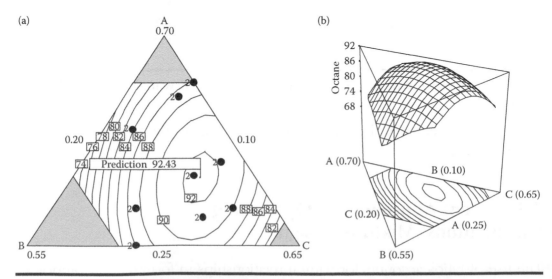

Figure 11.6 (a) Trilinear contour plot for the gasoline case. (b) 3D surface for octane.

be careful to put too much stock in the linear coefficients when you constrain the ingredients. For example, in this case, the predicted value of 56.23 (the coefficient for B in the equation) is an extrapolation for the octane of the purest B—which, as you can see from the contour plot in Figure 11.6a, reaches a theoretical value of 0.55, but in actuality, B never exceeded 0.4 in the blending experiment. Remember the mantra of DOE: Never extrapolate!

The optimum blend for octane is flagged on the contour plot in Figure 11.6a. It comes out to nearly the same composition (A = 0.4, B = 0.19, and C = 0.41) when predicted from the mixture model as it did from the original layout in ratio space. Whichever point of view is taken, the peak falls well within the explored space. However, as we've discussed, it's obvious from the plotted design points that the experiment laid out via ratios did a poor job exploring the extreme compositions that were considered feasible for blending.

DERIVATION OF SECOND-ORDER SCHEFFÉ POLYNOMIAL

Here is the derivation of the second-order Scheffé polynomials for two components. It takes the inherent constraint of mixtures, that is, $x_1 + x_2 = 1$, into account.

$\hat{y} = \beta_0 + \beta_1 x_1 + \beta_2 x_2 + \beta_{12} x_1 x_2 + \beta_{11} x_1^2 + \beta_{22} x_2^2$

Replace β_0 by $\beta_0(x_1 + x_2)$; x_1^2 by $x_1(1 - x_2)$; x_2^2 by $x_2(1 - x_1)$:

$\hat{y} = \beta_0(x_1 + x_2) + \beta_1 x_1 + \beta_2 x_2 + \beta_{12} x_1 x_2 + \beta_{11}(x_1(1 - x_2))$

$\quad\quad + \beta_{22}(x_2(1 - x_1))$

$\hat{y} = (\beta_0 + \beta_1 + \beta_{11})x_1 + (\beta_0 + \beta_2 + \beta_{22})x_2 + (\beta_{12} - \beta_{11} - \beta_{22})x_1 x_2$

$\hat{y} = \beta_1' x_1 + \beta_2' x_2 + \beta_{12}' x_1 x_2$

where $\beta_1' = \beta_0 + \beta_1 + \beta_{11}$, $\beta_2' = \beta_0 + \beta_2 + \beta_{22}$ and $\beta_{12}' = \beta_{12} - \beta_{11} - \beta_{22}$

These models, geared to mixtures, are distinguished by their lack of intercept. (The Scheffé coefficients incorporate the intercept [β_0] from the original equation.) What is the meaning of an intercept in mixtures? It would be the response when all the components are 0—this can't exist!

The gasoline-blending case presents an ideal application for an optimal design along the lines discussed in Chapter 7, where we introduced complex constraints as an aspect of RSM. Aided by Design-Expert software, we laid out the optimal mixture design shown in Figure 11.7. It's geared to fit a quadratic Scheffé polynomial. We then augmented the base optimal design (the six black circles) with three additional unique blends (open circles numbered in order of being picked) to test for LOF and match up with the original case that features nine compositions. This latter set of runs, called "check blends," is picked via the distance-based criterion. The remaining blank circles show candidate points that did not get chosen by either the optimal or the distance-based augmentation.

All nine of the chosen points could be replicated for the estimation of pure error, the same as before. At the very least, we'd recommend that the four most extreme vertices be replicated at random intervals in the blending runs. Another good candidate for replication would be the point in the middle—called the centroid in the jargon of the mixture design.

We've only scratched the surface of the mixture design. For more detail, see the two referenced texts by Cornell (2002) and Smith (2005). As you will see in Problem 11.2, setting up formulation problems via the tools of the mixture design is much more straightforward than taking the ratio route. These designs, being tailored to the mixture space, include more

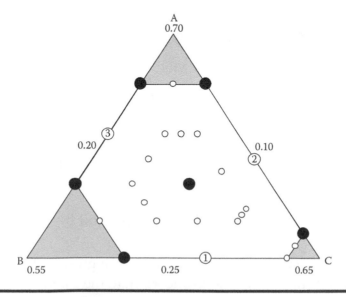

Figure 11.7 An alternative RSM design based on an optimal criterion for mixture space.

extreme compositions, that is, they are space filling, thus generating bigger effects that are more likely to emerge as significant signals. The use of Scheffé polynomials, the standard model for the mixture design, facilitates the interpretation of the component effects and interactions. If you get involved in formulation work, we urge you to look into this powerful tool for RSM.

PRACTICE PROBLEMS

11.1 You will be glad that we left this problem for last because it's a lollapalooza! A flexible part manufactured for medical use is made from four primary components:

- Resin A
- Crosslinker B
- Polymer X
- Polymer Y

The recipe for making this part is laid out as follows with the acceptable ranges listed:

1. Resin: 35–50 wt.% of copolymers (X and Y)
2. Crosslinker: 10–15 wt.% of copolymers
3. Polymer ratio: 60/40–80/20 X to Y

The response is elongation—the higher the better. A BBD will be done to optimize the formulation on the basis of this response. But first, some work must be done to set up proper ratios and translate them back to an actual composition.

The ratios can be defined as follows:

- $R_1 = A/(X + Y)$
- $R_2 = B/(X + Y)$
- $R_3 = X/Y$

Table 11.3 shows the ranges of the ratios to be studied via the BBD. For reasons described in the sidebar titled "Dealing with Nonlinear Behavior of Ratios," it will be laid out in terms of natural logarithms (shown in parentheses).

The resulting BBD is shown in Table 11.4. We translated back from the log scale to the original ratios by taking antilogs. These will then be converted into compositions for experimental purposes. However, given the responses for elongation listed in Table 11.4, along with the layout of inputs in log scale, you can develop a predictive model and perform the optimization (maximize).

Table 11.3   Ratio Constraints

Ratio	Description	Ratio Range	Low – (ln)	High + (ln)
R_1	Resin A as percent of the copolymer	35%–50%	0.35 (–1.050)	0.5 (–0.693)
R_2	Crosslinker B as percent of the copolymer	10%–15%	0.10 (–2.303)	0.15 (–1.897)
R_3	Polymer X to polymer Y	60/40–80/20	1.5 (0.405)	4.0 (1.386)

The mixture constraint is

■ $A + B + X + Y = 1$

From this and the ratio equations, the actual composition of the polymer is derived as follows:

■ $A = R_1/(R_1 + R_2 + 1)$
■ $B = R_2/(R_1 + R_2 + 1)$
■ $X = R_3/(R_3 + 1)(R_1 + R_2 + 1)$
■ $Y = 1/(R_3 + 1)(R_1 + R_2 + 1)$

Table 11.5 shows the compositions based on the ratios from Table 11.4 for the BBD. This is necessary for carrying out the experiment.

Ultimately you must translate the optimum point predicted from your model back to a composition by going through the same process detailed in Table 11.5:

1. Antilog each of the factor levels to translate them into actual ratios
2. Plug and chug these through the ratio equations to solve for A, B, X, and Y, the resin, crosslinker, and two polymers, respectively.

We never said this problem would be easy!

11.2 If you are not up to going through all the gyrations of applying RSM to formulations via the use of ratios, consider doing it in a more straightforward manner via the mixture design. From the website for the program associated with this book, follow the "Mixture Designs" link to a tutorial that provides an introduction to statistical tools for formulation developers. It details a case study on a detergent for which two responses were deemed to be the most important:

1. Viscosity
2. Turbidity

Table 11.4 BBD on Formulation for Medical Part

#	Coded			Actuals (Done in Natural Log Scale)			Elong. %
	R_1	R_2	R_3	R_1	R_2	R_3	
1	−1	−1	0	−1.050	−2.303	0.896	150
2	1	−1	0	−0.693	−2.303	0.896	164
3	−1	1	0	−1.050	−1.897	0.896	93.7
4	1	1	0	−0.693	−1.897	0.896	129
5	−1	0	−1	−1.050	−2.100	0.405	147
6	1	0	−1	−0.693	−2.100	0.405	175
7	−1	0	1	−1.050	−2.100	1.386	181
8	1	0	1	−0.693	−2.100	1.386	220
9	0	−1	−1	−0.872	−2.303	0.405	145
10	0	1	−1	−0.872	−1.897	0.405	128
11	0	−1	1	−0.872	−2.303	1.386	206
12	0	1	1	−0.872	−1.897	1.386	154
13	0	0	0	−0.872	−2.100	0.896	149
14	0	0	0	−0.872	−2.100	0.896	155
15	0	0	0	−0.872	−2.100	0.896	149
16	0	0	0	−0.872	−2.100	0.896	152
17	0	0	0	−0.872	−2.100	0.896	148

The formulators varied three components as shown below
- 3% ≤ A (water) ≤ 8%
- 2% ≤ B (alcohol) ≤ 4%
- 2% ≤ C (urea) ≤ 4%

They required that these three active components always equal 9 wt.% of the total formulation, that is,

$$A + B + C = 9\%$$

The other components (held constant) then must equal 91 wt.% of the detergent. Table 11.6 shows the experimental recipe sheet.

Table 11.5 Compositions for Experiment on Medical Part

#	Actuals (Antilogged)			Composition				Total
	R_1	R_2	R_3	A	B	C	D	
1	0.35	0.10	2.45	0.241	0.069	0.490	0.200	1.0
2	0.50	0.10	2.45	0.313	0.063	0.444	0.181	1.0
3	0.35	0.15	2.45	0.233	0.100	0.473	0.193	1.0
4	0.50	0.15	2.45	0.303	0.091	0.430	0.176	1.0
5	0.35	0.122	1.50	0.238	0.083	0.408	0.272	1.0
6	0.50	0.122	1.50	0.308	0.075	0.370	0.247	1.0
7	0.35	0.122	4.00	0.238	0.083	0.543	0.136	1.0
8	0.50	0.122	4.00	0.308	0.075	0.493	0.123	1.0
9	0.418	0.10	1.50	0.275	0.066	0.395	0.264	1.0
10	0.418	0.15	1.50	0.267	0.096	0.383	0.255	1.0
11	0.418	0.10	4.00	0.275	0.066	0.527	0.132	1.0
12	0.418	0.15	4.00	0.267	0.096	0.510	0.128	1.0
13	0.418	0.122	2.45	0.271	0.079	0.461	0.188	1.0
14	0.418	0.122	2.45	0.271	0.079	0.461	0.188	1.0
15	0.418	0.122	2.45	0.271	0.079	0.461	0.188	1.0
16	0.418	0.122	2.45	0.271	0.079	0.461	0.188	1.0
17	0.418	0.122	2.45	0.271	0.079	0.461	0.188	1.0

Note that the last four runs in standard (Std) order are replicates of the three vertices in the triangular mixture space (a simplex) and the overall centroid (the equivalent of a CP in an RSM design). You will see the design pictured in the software tutorial (and perhaps on screen as well, if you bring up the program on your computer). We dare not get into any more detail—that must await another book devoted to simplifying the mixture design. The first several chapters have been written and posted as a primer posted at www.stateease.com/formulator. Check it out!

Table 11.6 Mixture Design for a Detergent

Std	Type	A: Water	B: Alcohol	C: Urea	Viscosity mPa-sec	Turbidity
1	Vertex	5.00	2.00	2.00	40.8	436
2	Edge	4.00	3.00	2.00	67.9	436
3	Edge	4.00	2.00	3.00	46.5	630
4	Vertex	3.00	4.00	2.00	87.8	323
5	Edge	3.00	3.00	3.00	45.3	949
6	Vertex	3.00	2.00	4.00	144	641
7	Check	4.33	2.33	2.33	35.1	671
8	Check	3.33	3.33	2.33	51.7	730
9	Check	3.33	2.33	3.33	70.7	874
10	Centroid	3.67	2.67	2.67	46	1122
11	Vertex	5.00	2.00	2.00	37.2	378
12	Vertex	3.00	2.00	4.00	130	786
13	Vertex	3.00	4.00	2.00	91.6	546
14	Centroid	3.67	2.67	2.67	34.8	984

Chapter 12

Practical Aspects for RSM Success

Keep on going and the chances are you will stumble on something, perhaps when you are least expecting it. I have never heard of anyone stumbling on something sitting down.

Charles Kettering
American engineer and the holder of 186 patents

In this last chapter, you may very well stumble across a practical aspect on RSM that will put you over the top for your optimization goals. Here, we finish off our book with vital tools for handling hard-to-change (HTC) factors, right sizing your RSM design, and how to confirm your outstanding results.

Split Plots to Handle HTC Factors

As we detailed in the third edition of *DOE Simplified* (2015) in a new chapter (11) devoted to this topic, experimental factors often cannot be randomized easily. Split plot designs then become attractive by grouping HTC factors, albeit with a loss in power (Anderson and Whitcomb, 2014). While split plots for factorial experiments go back to nearly a century (Fisher, 1925), application of these restricted-randomization designs to RSM is a relatively recent development, finding use, for example, in wind tunnel testing, where some factors, such as wing-tip height, cannot be changed easily (English, 2007).

The details on the design and analysis of RSM split plot experiments goes beyond the scope of this book (refer to Vining et al., 2005). However, to provide an idea of how such a design comes together, let's go back to the trebuchet case from Chapter 5. There, we reported the results of an experiment on three factors—arm length, counterweight, and missile weight—on projectile distance. It turns out, as reported by the PBS show Nova (www.pbs.org/wgbh/nova/lostempires/trebuchet/wheels.html), that "trebs" on wheels fire a lot further than the ones that remain stationary. This is purely a matter of physics as we spelled out in our note on the "Physics of the trebuchet" in Chapter 5.

**MARCO POLO MISSES THE MARK BY NOT
PUTTING WHEELS ON HIS TREBUCHET**

The producers of the 2014 Netflix/Weinstein series *Marco Polo* built trebuchets that threw a 25-kilogram sandbag almost 300 meters. As you can see from the YouTube video posted at www.youtube.com/watch?v=F8xW-LkFq6A, these were not mounted on wheels, which might have added another 100 meters (33%) to the distance according to one source (www.midi-france.info/medievalwarfare/121343_perriers.htm).

Consider a "thought experiment" on the machine pictured in Figure 5.2 where we add a fourth factor—wheels on or off. It would be very inconvenient to randomly install these or remove them between each throw. Thus, a split plot that groups firings of the trebuchet with or without wheels would be a very convenient way to accomplish this experiment. Table 12.1 lays out an optimal design with wheels being the grouped ("Grp") HTC factor "d" (changing the case to differentiate from the easy-to-change factors randomized within groups).

By its grouping on column "d," this 32-run split plot design provides a practical way for quantifying the effects of putting wheels on the trebuchet. Now, according to the randomized test plan, they only need to be changed every four runs—not one by one.

Table 12.1   Optimal Split Plot Design on Trebuchet

Grp	Run	A	B	C	d	y
1	1	4	10	2.5	On	
1	2	4	20	2	On	
1	3	8	15	2	On	
1	4	6	20	3	On	
2	5	4	10	3	On	
2	6	8	20	2.5	On	
2	7	6	10	2	On	
2	8	8	15	3	On	
3	9	8	10	3	On	
3	10	4	20	3	On	
3	11	8	20	2	On	
3	12	4	15	2	On	
4	13	8	10	3	Off	
4	14	4	20	2.5	Off	
4	15	4	10	2	Off	
4	16	8	20	2	Off	
5	17	6	15	2.5	On	
5	18	8	20	3	On	
5	19	8	10	2	On	
5	20	4	20	2	On	
6	21	4	20	3	Off	
6	22	8	15	2	Off	
6	23	6	10	3	Off	
6	24	4	10	2	Off	
7	25	4	10	3	Off	
7	26	6	20	2	Off	
7	27	8	20	3	Off	

(Continued)

Table 12.1 (*Continued*) Optimal Split Plot Design on Trebuchet

Grp	Run	A	B	C	d	y
7	28	8	10	2	Off	
8	29	4	15	3	Off	
8	30	8	10	2.5	Off	
8	31	4	20	2	Off	
8	32	8	20	3	Off	

RESET HTC FACTOR(S) BETWEEN GROUPS!

When you hit the double lines drawn between groups, stop everything and see what is called for next on the HTC factors. If any or all do not change, it will be tempting to just leave them as they are. However, to generate a true measure of error that includes variation from the setup, you must reset all HTC factors, for example, take the wheels off and put them back on.

Ideally, the reset should be done from the ground up. However, when this is completely impractical, do the best you can to break the continuity of the HTC setting. For example, for furnace temperature (a common HTC factor), it might take too long for you to let things cool back down to ambient temperature. In that case, at least, open the door, remove what was on it (such as your chocolate chip cookies), turn the dial (or reset the digital control) to another temperature, and then back to the upcoming group level (e.g., 350 degrees Fahrenheit or 175 degrees Celsius for the cookies). This can be considered sufficient to reset the system. But if you do so, note it in your report as a shortcut. Full disclosure on actual experimental procedures is vital for an accurate interpretation of results by statisticians and subject-matter experts, as well.

Right-Sizing Designs via Fraction of Design Space Plots

By sizing experiment designs properly, researchers can assure that they specify a sufficient number of runs to reveal any important effects on the system. For factorial designs, this can be done by calculating statistical power (Anderson and Whitcomb, March 2014). However, the test matrices for RSM generally do not exhibit orthogonality; thus, the effect on calculations becomes correlated and degrades the statistical power. This becomes especially troublesome for constrained designs such as the one for the experiment on the air-driven mechanical part that we illustrated in Figure 7.6.

Fortunately, there is a work-around to using power to size designs that makes use of a tool called "fraction of design space" (FDS) plot (Zahran et al., 2003). The FDS plot displays the proportion (from zero to one) of the experiment-design space falling below the PV value (specified in terms of the standard error (SE) of the mean) on the y-axis.

THREE THINGS THAT AFFECT PRECISION OF PREDICTION

When the goal is optimization (usually the case for RSM), the emphasis is on producing a precisely fitted surface. The response surface is drawn by predicting the mean outcome as a function of inputs over the region of experimentation. How precisely the surface can be fit (or the mean values estimated) is a function of the SE of the predicted mean response—the smaller the SE the better. The SE of the predicted mean response at any point in the design space is a function of three things:

1. The experimental error (expressed as a standard deviation)
2. The experiment design—then the number of runs and their location
3. Where the point is located in the design space (its coordinates)

To keep things really simple for illustrating FDS plots, let's consider fitting a straight line, the thick one shown in Figure 12.1, to response data (y) as a function of one factor (x). Naturally, as more runs are made, the more precise this fit becomes. However, due to practical limits on the run budget, it becomes necessary to be realistic in establishing the desired precision "d"—the half-width of the thin-lined interval depicted in Figure 12.1.

This simple case simulates five runs—two at each end and one in the middle of the 0–1 range—from which a CI can be calculated. The CI is displayed by the dotted lines that flare out characteristically at the extremes of the factor. Unfortunately, only 51% of the experimental region provides mean response predictions within the desired precision plus-or-minus d. This is the FDS, reported as 0.51 on a scale of zero to one. More runs, only a few in this case, are needed to push FDS above the generally acceptable level of 80%.

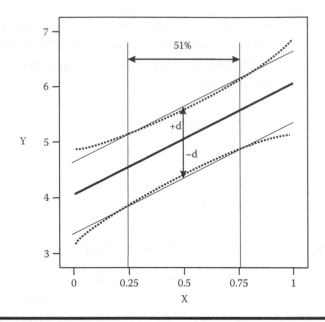

Figure 12.1   FDS illustrated for a simple one-factor experiment.

Now that you have seen what FDS measures for one factor, let's move up by 1D to a 2D view by revisiting the case of the air-driven mechanical part. Figure 12.2 illustrates an FDS graph with the crosshair set at the level of 0.5 on the SE of the mean (y-axis). It tells us that about one-third (FDS = 0.34) of the experimental region will provide predictions within this specified SE with 95% confidence (alpha of 0.05).

Figure 12.3, a contour plot of SE, shows which regions fall below the 0.5 level of SE.

Notice that only about a third of the triangular experimental region falls within the two areas bordered by the 0.5 SE contours. This directly corresponds to the 0.34 FDS measure. Again, look at Figure 12.3 and find the 0.6 SE contour—it falls near the perimeter of the experimental region. The FDS for 0.6 SE comes to 0.83; in other words, 83% of the experimental region can be estimated to within this level of precision.

Whether the 15-run design (the points in Figure 12.3, three of which are replicated) is sized properly depends on the ratio of the required precision "d" to the standard deviation "s." Given software that generates FDS plots (such as the program accompanying this book), you need to only enter these two numbers to evaluate your design. In this case, let's assume that the values for d and s are 1.5 and 1, respectively. The SE derived from this 1.5 standard-deviation requirement on precision is 0.663. (To be specific,

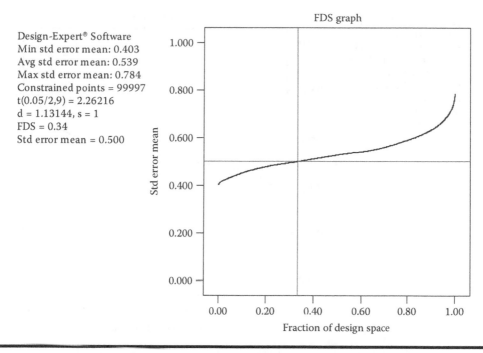

Figure 12.2   FDS illustrated for the two-factor experiment on an air-driven mechanical part.

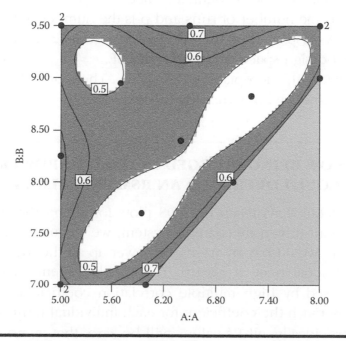

Figure 12.3   Contour plot of SE with regions above 0.5 shaded out.

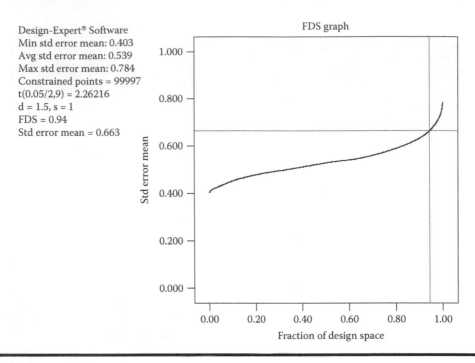

Design-Expert® Software
Min std error mean: 0.403
Avg std error mean: 0.539
Max std error mean: 0.784
Constrained points = 99997
t(0.05/2,9) = 2.26216
d = 1.5, s = 1
FDS = 0.94
Std error mean = 0.663

Figure 12.4   FDS based on actual requirements for RSM experiment on the air-driven mechanical system.

it equals 1.5 divided by the two-tailed t-value for alpha of 0.05 with N minus p df, where N is the number of runs and p is the number of terms including the intercept.) As displayed by the FDS plot in Figure 12.4, a healthy 94% of the experiment-design space falls within this SE.

This far exceeds the acceptable FDS of 80%. Therefore, all systems are going to the engineer who is running this experiment.

HOW BEING PLUGGED IN ONLY TO POWER COULD OVERLOAD AN RSM EXPERIMENT

In a more detailed write-up on FDS plots for sizing the RSM experiment on the air-driven mechanical system, we calculate the price that test engineers would pay by trying to power up all the individual effect estimates (Anderson et al., 2016). The problem stems from variance inflation caused by high multiple correlation coefficients (R_i^2). These indicate how much the coefficient for each individual term is correlated to the others. Ideally, all R_i^2 values will be zero, that is, no correlation, and thus the matrix becomes orthogonal. However, this cannot be counted on for RSM experiments. In this case, terms A, B, and AB come in at 0.778, 0.764, and 0.783 multiple correlation coefficients, respectively. To get the worst of these three estimates, AB by a hair, powered up properly to 80% for detecting a 1.5 standard-deviation effect would require more than 200 test runs. That would sink the experiment unnecessarily—given that only 15 runs suffice on the basis of the FDS sizing, as we just illustrated. So, you would at best unplug the power for assessing RSM designs.

For nonorthogonal designs, such as RSM along the lines of our example, FDS plotting works much better and more appropriately for the purpose of optimization than statistical power for "right-sizing" experiment designs. They generally lead to a smaller number of runs while providing the assurance needed that, whether or not anything emerges to be significant, the results will be statistically defensible.

How to Confirm Your Models

Process modeling via RSM relies on empirical data and approximation by polynomial equations. As we spelled out in Chapter 2, the end result may be of no use whatsoever for prediction. Therefore, it is essential that all RSM models must be confirmed.

One approach for confirmation, which we discussed in Chapter 2, randomly splits out a subset, up to 50%, of the raw data into a validation sample. However, this works well only for datasets above a certain size (refer back to a formulaic rule of thumb) as it wastes too many runs (50% if you follow the cited protocol). A far simpler and most obvious way to confirm a model is to follow-up your experiment with one or more runs at the recommended conditions. Then, see if they fall within the 95% PI, which we detailed in Chapter 4 in the note on "Formulas for Standard Error (SE) of Predictions."

To provide a convincing evidence of confirmation, you should replicate the confirmation run at least three times. More runs are better. However, your returns will diminish due to the PI being bounded by the CI, which is dictated by the sample size N—the number of runs in your original experiment. For example, going back to the chemical reaction case in Chapter 4, we reported that the optimal conditions produced a predicted yield of 89.1 grams with a 95% PI of 85.4–92.9 grams for a single-confirmation run (n = 1). Here is the progression of this PI for further confirmation runs:

■ Three (n = 3): 86.3–92.0 grams
■ Six (6): 86.5–91.8 grams
■ Twelve (12): 86.7–91.6 grams
■ One million (1,000,000): 86.8–91.5 grams

The 95% CI is 86.8–91.5 grams. These are the limiting PI values as you can surmise from the interval for a million runs.

Figure 12.5 graphically lays out the probability of detecting an issue versus the number of confirmation runs (Bezener, 2015).

We recommend that at the least, you do two confirmation runs at the predicted optimum, but six or so would be nearly ideal. The best return on investment in confirmation comes with 5–10 runs. Beyond 10 runs, the investment in confirmation does not provide a good return (Jensen, 2015).

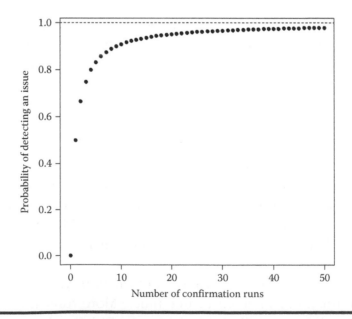

Figure 12.5 Probability of detecting an issue versus the number of confirmation runs.

PERFECTING THE RECIPE FOR RAMEN NOODLES: CASE FOR CONFIRMATION

Programmers and statisticians at Stat-Ease teamed up on an experiment to optimize the recipe for ramen noodles as a function of the amount of water, cooking time, brand, and flavor. They settled on chicken that produced great sensory results no matter what the brand, provided it was produced with just the right amount of water and cooked for an optimal time. The lead experimenter, Brooks Henderson, then produced seven bowls of ramen, three according to the perfect recipe, and four others at inferior conditions to keep the tasters honest. Happily, everything fell into place for all the measured attributes, which included not only the sensory evaluations for taste and crunchiness, but also physical changes caused by the hydration process. For example, the model predicted a weight gain of 174% for the noodles. The confirmation average (n = 3) came only to 160.2%. However, this fell well within the 95% PI of 137.3%–210.9%. This is the way to fuel technical workers, that is, provide good-quality ramen that soaks up lots of water and therefore maximizes the well-being that results from a full stomach.

Confirming the Optimal Ramen, Stat-Teaser Newsletter,
January 2013, p. 3, Stat-Ease, Inc.

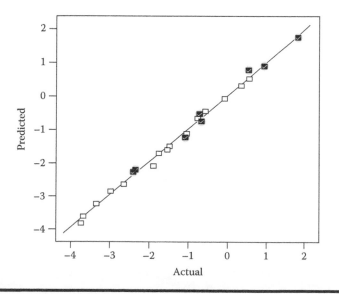

Figure 12.6 **Actual versus predicted response plot included with verification points.**

Another approach for confirmation is to do it concurrently, that is, during the actual experiment. This makes sense when it takes a long time to complete a block of runs and/or measure the responses. Sometimes, the opportunity for experimentation, for example, on a manufacturing scale, is limited to one shot. In these situations, it will be advantageous to embed a number of runs within your experiment design that are set aside from the others only for verification. These verification runs are added to and randomized with the DOE runs. However, they are not used to estimate the model; only the design runs are used. Then, after the experiment is completed and the results are analyzed, the predicted values of the verification runs are compared to their observed values. If they fall in line, then the model is confirmed.

Figure 12.6 shows an example where the verification points fall in line with their predicted values.

In this case, the model is clearly confirmed.

On that positive note, we conclude our primer on RSM. We hope that you will do well by applying RSM to your process—all the best!

Glossary

Statistical Symbols

d_i individual desirability

D   Cook's distance; overall desirability

e   raw residual (or error)

F   F statistic (a ratio of two mean squares [MSs] named after Sir Ronald Fisher)

h_{ij} elements of the hat matrix **H**

H   hat matrix

k   number of factors in the design

n   number of observations in a sample

p   fraction of the design (e.g., 2^{k-p}); probability value ("p-value"); number of model parameters including the intercept and block; coefficients

r   sample correlation coefficient

R^2 coefficient of determination

s   sample standard deviation

s^2 sample variance

t   t-value

V   variance

x   independent variable

y   observed response value

X   matrix of independent variables (observed values of predictors)

Y   matrix of observed response values

α ("alpha") coded-unit distance for axial points in the central composite design (CCD) β   ("beta") coefficient in the predictive model

Δ   ("delta") change (e.g., Δx)

∇   ("nabla") matrix of partial derivatives (e.g., $\nabla = \begin{pmatrix} \dfrac{\partial f}{\partial X_1} \\ \dfrac{\partial f}{\partial X_2} \\ \dfrac{\partial f}{\partial X_k} \end{pmatrix}$)

λ   ("lambda") power transformation (e.g., y^λ)

Π   (capital "pi") mathematical operator to take the product of a number series

σ   ("sigma") population (true) standard deviation

Σ   (capital sigma) mathematical operator to take the sum of a number series variance matrix (e.g., $\Sigma_{adj} = \begin{pmatrix} \sigma_{11}^2 & \sigma_{12}^2 & \sigma_{1k}^2 \\ \sigma_{12}^2 & \sigma_{22}^2 & \sigma_{2k}^2 \\ \sigma_{1k}^2 & \sigma_{2k}^2 & \sigma_{kk}^2 \end{pmatrix}$)

Subscripts

i   individual datum

i − 1   all datum except individual i

j   other individual datum, not i

Superscripts

*   (star) of a given value (e.g., x*)

^   (hat) predicted

—   (bar) average

′   (prime) transpose of the design matrix (sometimes symbolized as T)

−1   inverse

Terms

Actual value: The observed value of the response from the experiment. The physical levels of the variables in their units of measure (as opposed to their coded levels, such as −1 or +1).

Adequate precision: A measure of the experimental signal-to-noise ratio. Ratios greater than four are good.

$$\left[\frac{\max(\hat{y}) - \min(\hat{y})}{\sqrt{\bar{V}(\hat{y})}} \right] \bar{V}(\hat{y}) = \frac{1}{n} \sum_{i=1}^{n} V(\hat{y}) = \frac{p\sigma^2}{n}$$

It compares the range of the predicted values, $\hat{y}$, at the design points to the average variance, V-bar, of the prediction (a function of model parameters, p, the number of points, n, and the variance, σ^2, estimated by the root MS residual from ANOVA).

Adjusted R-squared: R-squared adjusted for the number of terms in the model relative to the number of points in the design. An estimate of the fraction of overall variation in the data accounted for by the model.

Akaike information criterion corrected (AICc): A way to compare different models on a given outcome that balances underfitting a model, which may not capture the true nature of the variability in the outcome variable, versus an overfitted model that loses generality. The lower the AICc, the better the model. The AICc together with forward selection is favored for model selection from supersatured experiments such as definitive-screening designs.

Alias: Other term(s) that is (are) correlated with a given coefficient. The resulting predictive model is then said to be "aliased." (It is also called "confounding.")

AN EXTREME EXAMPLE OF ALIASING

As discussed in Chapter 3 (sidebar "Ladies Tasting Tea versus Girls Gulping Soda"), Mark's youngest daughter Katie (shown in the picture) challenged him to blind a taste test of Coca Cola® (aka Coke) versus Pepsi Cola® (aka Pepsi) soft drinks. Her seventh-grade social studies teacher had done something similar in her class that day to illustrate how advertising influences consumer perceptions. Mark was tired after a day of teaching design of experiments and figured out that a good dose of caffeine might help him do his moonlighting job as a primary author of *RSM Simplified*. Imagine his shock when Katie paraded in with two glasses—one made of white foam and the other made from blue plastic! She explained that, although Mark, and others in the household to be tested, should not be told which cola was which, Katie needed to keep them straight. Unfortunately, by setting up her experiment in this manner,

Katie aliased the effect of Pepsi versus Coke with the bias drinkers might have between cup types (for example, some people dislike plastic and others believe that foam may be bad for the environment).

ANOVA: A statistical method based on the F-test, which assesses the significance of experimental results. It involves subdividing the total variation of a set of data into component parts.

Antagonism: An undesirable interaction of two factors where the combination produces a response that is not as good as what would be expected from either one alone. The same concept can be applied to higher-order interactions.

Average: See "Mean."

Axial points: Design points that fall on the spatial coordinate axes emanating from the overall center point (or centroid in mixture space), often used as a label for star points in a CCD.

Bias: A systematic error in the estimation of a population value.

Block: A group of trials based on a common factor. Blocking is advantageous when there is a known factor that may influence the experimental result, but the effect is not of interest. For example, if all runs cannot be conducted in 1 day or within one batch of raw material, the experimental points can be divided in such a way that the blocked effect is eliminated before computation of the model. Removal of the block effect reduces the noise in the experiment and improves the sensitivity to effects.

Box–Behnken design (BBD): An RSM design where each factor takes only three levels. Unlike composite designs, the BBD does not embed a full or fractional factorial design.

Box–Cox plot: A diagnostic tool for selecting transformations that display the scaled model residual sum of squares (SS) versus the power taken from the response.

Case statistics: Diagnostic statistics calculated for each run after the model has been selected.

Categoric variable: Factors whose levels fall into distinct nonnumeric classes, such as metal versus a plastic material. (It is also called a "class" or "qualitative" variable.)

Center point: A run with all numerical factor levels set at their midpoint value.

CCD: A design for RSM that's composed of a core two-level factorial plus axial points and center points. If not face centered, each factor has five levels. Be careful to be sure that your region of operability is greater than the region of interest to accommodate axial runs.

Centroid: The center point of mixture space within the specified constraints.

Check blend: A unique composition added to the experiment design, above and beyond the number of blends needed for the chosen polynomial model, to test for the lack of fit (LOF). These points are best placed in the gaps between model points.

Class variable: See "Categoric variable."

Coded-factor level: See "Coding."

Coding: A way to center and normalize factors, for example, by converting low- and high-factor levels into −1 and +1, respectively.

Coefficient: See "Model coefficient."

Coefficient of variation (CV): Also known as the "relative standard deviation," the CV is a measure of residual variation of the data relative to the size of the mean. It is the standard deviation (root MS error from ANOVA) divided by the dependent mean, expressed as a percent.

Component: An ingredient of a mixture.

Confidence interval: A data-based interval constructed by a method that covers the true population parameter at a stated percentage (typically 95%) of the time in repeated samplings.

Confounding: See "Alias."

Constraint: Limit in respect to factor levels in a process design or component ranges for a mixture experiment.

Continuous variable: See "Numerical variable."

Contour plot: A topographical map drawn from a mathematical model, usually in conjunction with RSM for an experimental design. Each contour represents a continuous response fixed at some value.

Convex hull: A geometric shape that can be imagined as a rubber band that's been stretched around a set of points and then released.

The resulting region will conform as closely as possible to the points, but with no indentations. This geometry is useful for defining constraints. It enables picking new points from existing combinations, such as the midpoint of two vertices. If the hull is not convex, such points may be falsely rejected for being out of bounds.

Cook's distance: A measure of how much the fitted model would change if a particular run were omitted from the analysis. Relatively large values are associated with cases with a high leverage and large studentized residuals. Cases with large values relative to the other cases may cause undue influence on the fitting and should be investigated. They could be caused by recording errors, an incorrect model, or a design point far from the remaining cases.

Corrected total: The total SS corrected for the mean (calculated by taking the sum of the squared distances of each individual response value from its overall average).

Cubic model: A polynomial equation used for predictive purposes that contains terms of first order (linear or main effects), second order (two-factor interaction and squared terms), and third order (three-factor interactions, cubed terms, and everything in-between).

Cumulative probability: The proportion of individuals in a population that fall below a specified value.

Curvature: A measure of the offset at the center point of actual versus predicted values from a two-level factorial model. If significant, consider augmenting to a quadratic model, which can be fit to data from a response surface design.

Degree of equation: The highest order of terms in a model. For example, in an equation of degree two, you will find terms with two factors multiplied together as well as squared terms.

Degrees of freedom (df): The number of independent comparisons available to estimate a parameter.

Dependent mean: The mean of the response over all the design points.

Design matrix: An array of values presented in rows and columns. The columns usually represent design factors. The values in the rows represent settings for each factor in the individual runs of the design.

Design parameters: The number of levels, factors, replicates, and blocks within the design.

Design of experiment space: An imaginary area bounded by the extremes of the tested factors. (It is also called an "experimental region.")

Desirability: A utility function that ranges from zero (not acceptable) to one (ideal), which makes it possible to optimize multiple responses simultaneously via numerical methods.

Deterministic: An outcome that does not vary (i.e., it is always the same) for a given set of input factors.

DFBETAS: A measure of influence based on the difference in betas (model coefficients) that occurs when that run is deleted.

DFFITS: A measure of influence based on the difference in fits in each predicted value that occurs when a particular run is deleted.

Diagnostics: Statistics and plots, often involving model residuals, which assess the assumptions underlying a statistical analysis.

Distribution: A spatial array of data values.

D-optimal: A criterion for choosing design points via minimization of the determinant of the $(\mathbf{X'X})^{-1}$ matrix. It minimizes the volume of the confidence ellipsoid for the coefficients, thereby providing the most precise coefficient estimates possible. Equivalently, D-optimality maximizes the determinant $(\mathbf{X'X})$.

D-optimal design: An experimental design that chooses runs, often from a larger candidate set of points, based on the D-optimal criterion.

Effect: The change in average response when a factor, or interaction of factors, goes from its low level to its high level.

Error term: The term in the model that represents random error. The data residuals are used to estimate the nature of the error term. The usual assumption is that the error term is normally and randomly distributed about zero, with a constant standard deviation of sigma.

Euclidean norm: A distance calculated by summing up squares of all coordinates and taking the square root; in essence Pythagorus's theorem.

EVOP: Evolutionary operation, an ongoing method for experimenting on a full-scale process, with minimum disruption, so that information on how to improve the process is generated.

Experiment: A series of test runs for the purpose of discovery.

Experimental region: See "Design of experiment space."

Externally studentized residual: (Also see "outlier t" and "R-student.") This statistic tests whether a run is consistent with other runs, assuming that the chosen model holds good. Model coefficients are calculated based on all design points except one. A prediction of the response at this point is then produced. The externally studentized residual measures the number of standard deviations difference between this new predicted value (lacking the point in question) and the actual response. Note: this statistic becomes undefined for points with leverages of one.

Face-centered CCD (FCD): A type of CCD that brings the star points to the faces of the factorial core (coded-distance alpha equals one). Each factor only has three levels. Use the FCD when your region of interest and region of operability coincide.

Factor: The independent variable to be manipulated in an experiment.

Factorial design: A series of runs, in which all combinations of factor levels are included. It can be "full" or "fractional."

F-distribution: A probability distribution used in the ANOVA. The F-distribution is dependent on the degrees of freedom (df) for the MS in the numerator and the df of the MS in the denominator of the F-ratio.

First-order model: A polynomial model containing only linear or main-effect terms.

Fisher's information matrix: A calculation made to evaluate a design matrix that is defined as $\mathbf{X'X}$ divided by the variance (set at one for design evaluation purposes).

Fractional factorial: An experimental design including only a subset of all possible combinations of factor levels, causing some of the effects to be aliased.

Fraction of design space plot: A tool for sizing designs that graphs the proportion (from zero to one) of the experimental region falling below the prediction variance (PV) value specified on the y-axis.

F-test: See "F-value."

Full factorial: An experimental design including all possible combinations of factors at their designated levels.

F-value: The F-distribution is a probability distribution used to compare variances by examining their ratio. If they are equal, the F-value is 1. The F-value in the ANOVA table is the ratio of the model MS to the appropriate error MS. The larger their ratio, the larger the F-value and the more likely that the variance contributed by the model is significantly larger than random error. (It is also called the "F-test.")

Happenstance data: Information from historical records as opposed to a designed experiment. Typically, the response variation will be largely due to noise in the system and little useful statistical information may be obtained. Cause-and-effect relations cannot be reliably obtained by regression of such data.

Hat matrix (H): A matrix calculated from input values (x) defined as

$$H = X(X'X)^{-1}X' = [h_{ij}]$$

where **X** is a design matrix of n rows and p columns and **H** is an n-by-n symmetric matrix. It converts the matrix of response values (**Y**) into the predicted outcomes ($\hat{Y}$), that is, $\hat{Y} = HY$.

Hierarchy: The ancestral lineage of effects flowing from the main effects (parents) down through successive generations of higher-order interactions (children). For statistical reasons, models containing subsets of all possible effects should preserve hierarchy. Although the response may be predicted without the main effects when using the coded variables, predictions will not be the same in the actual factor levels, unless the main effects are included in the model. Without the main effects, the model will be scale dependent.

Hybrid design: A minimum-run RSM design created by applying a CCD for all factors except the last, which gets chosen D-optimally. This design exhibits poor design properties and it is very sensitive to outliers and missing data. Each factor has four or five levels. The region of operability must be greater than the region of interest to accommodate axial runs.

Hypothesis (H): A mathematical proposition set forth as an explanation of a scientific phenomenon.

Hypothesis test: A statistical method to assess the consistency of the observed data with a stated hypothesis.

Idempotent: A matrix **A** for which $A^2 = A^*A = A$, that is, it is unchanged in value when multiplied or otherwise operated on by itself.

Identity column (I): A column of all plus 1s in the coded-factor design matrix, used to calculate the model intercept.

Independence: A desirable statistical property where knowing the outcome of one event tells nothing about what will happen from another event.

Individuals: Discrete subjects or data from the population, that is, experimental units.

Interaction: The combined change in two factors that produces an effect different from that of the sum of effects from the two factors. Interactions occur when the effect of one factor depends on the level of another factor.

Intercept: The constant in the regression equation.

Internally studentized residual: The residual divided by the estimated standard deviation of that residual. It is a function of leverages as shown by this equation

$$r_i = \frac{e_i}{s\sqrt{1 - h_{ii}}}$$

where e_i represents the raw residual (the difference between the actual and predicted response). In essence, the studentized residual measures the number of standard deviations separating the actual value from predicted values. Note that this statistic becomes undefined for cases with leverages of one.

I-optimal: A criterion for choosing design points via minimization of the integral of the PV across the design space. Also referred to as "IV."

Irregular fraction: A two-level fractional factorial design that contains the total number of runs that is not a power of two. For example, a 12-run fraction of the 16-run full factorial design on four factors. This is a 3/4 irregular fraction.

LOF: A test that compares the deviation of actual points from the fitted surface, relative to pure error. If a model has a significant LOF, it should be investigated before being used for prediction.

Latin square: A multiple blocked design involving factors, in which the combination of the levels of any one of them with the levels of the other two appears once and only once. It was first used in agricultural research to eliminate environmental variations, but was not used much in industrial experimentation.

Least-significant difference (LSD): A numerical value used as a benchmark for comparing treatment means. When the LSD is exceeded, the means are considered to be significantly different.

Least squares: See "Regression analysis."

Level: The setting of a factor.

Leverage: The potential for a design point to influence its fitted value, which is defined as the diagonal element h_{ii} of the **H** matrix. It represents the fraction of the error variance, associated with the point estimate, carried into the model. The sum of leverage values for all design points equals the number of coefficients (including the constant) fit by the predictive model. Leverages near one (the maximum) should be avoided because the model will then be forced to fit whatever response value is entered for that particular run (predicted equates to actual). Replication of such design points is advised to reduce their leverage. A design point can have the maximum leverage of 1/k, where k is the number of times the design point is replicated. A point with leverage greater than two times the average is generally regarded as having high leverage, in other words, an outlier in the independent variable space. Note that leverage only takes into account the spatial location of the point, not the data observed there.

Linear model: See "First-order model."

Logit transformation: A mathematical function defined as

$$\text{Logit}(y) = \log_e[(y - \text{lower limit})/(\text{upper limit} - y)]$$

It's used as a variance stabilizer for bounded data, for example, molecular yield ranging from a lower limit of 0 to an upper limit of 100%.

LSD bars: Plotted intervals around the means on effect graphs with lengths set at one-half of the LSD. Bars that do not overlap indicate significant pair-wise differences between specific treatments.

Lurking variable: An unobserved factor (one not in the design) causing a change in response. A classic example is the study of population in Oldenburg versus the number of storks, which led to a spurious conclusion that storks cause babies (Sies, 1988).

Main effect: The change in response caused by changing a single factor, holding all the others constant.

Mean: The sum of all data divided by the number of data—a measure of location (center). (It is also called "average.")

MS: The SS divided by its degrees of freedom (SS/df). It is an estimate of variance.

Median: The middle value.

Mixture model: See "Scheffé polynomial."

Model: An equation, typically a polynomial, which is fit to the data in an attempt to describe a physical process for the purpose of prediction.

Model coefficient: The coefficient of a factor in a model. (It is also called a "parameter" or "term.")

Multicollinearity: The problem of correlation of one variable with others that arise when the predictor variables are highly interrelated (i.e., some predictors are nearly linear combinations of others). Highly collinear models tend to have unstable regression coefficient estimates.

Multiple-response optimization: Method(s) for simultaneously finding the combination of factors giving the most desirable outcome for more than one response.

Noise: Uncontrolled source of variation.

Normal distribution: A distribution for variable data, represented by a bell-shaped curve symmetrical about the mean with a dispersion specified by its standard deviation.

Normal probability plot: A graph with a y-axis that is scaled by cumulative probability (Z) so that normal data plot as a straight line.

Null: Zero difference.

Numerical variable: A quantitative factor that can be varied on a continuous scale, such as temperature.

One-factor-at-a-time: One-factor-at-a-time method of experimentation.

Observation: A record of factor levels and associated responses for a particular experimental run (trial).

Order: A measure of complexity of a polynomial model. For example, first-order models contain only linear terms. Second-order models contain linear terms plus two-factor interaction terms and/or squared terms. The higher the order, the more complex the forms of curvature that the polynomial model can approximate.

Orthogonal arrays: Design matrices exhibiting the property of orthogonality.

Orthogonality: A property of a design matrix that exhibits no correlation among its factors, thus allowing them to be estimated independently.

Outlier: A design point where the response does not fit the model.

Outlier t-test: See "Externally studentized residual."

Parameter: See "Model coefficient."

Pencil test: A quick and dirty method for determining whether a series of points falls on a line. If they are covered, the points pass the test.

Perturbation plot: A graph showing how the response varies as a function of all input factors varied one at a time from a reference point (normally the center of the design space).

Plackett–Burman design: A class of orthogonal (for the main effects) fractional two-level factorial designs where the number of runs is a multiple of four, rather than 2^k. These designs are of Resolution III.

Polynomials: Mathematical expressions, composed of powers of predictors with various orders, used to approximate a true relationship.

Population: A finite or infinite collection of all possible individuals who share a defined characteristic, for example, all parts made by a specific process.

Power: The probability that a test will reveal an effect of a stated size, if it is actually present.

Power law: A relationship where one variable (e.g., standard deviation) is proportional to another variable (such as the mean) raised to a power.

Predicted R-squared: Measures the amount of variation in new data explained by the model. It makes use of the predicted residual sum of squares (PRESS) as shown in the following equation: Predicted R-squared = $1- SS_{PRESS}/(SS_{TOTAL}-SS_{BLOCKS})$.

Predicted value: The value of the response predicted by the mathematical model, for example, defined in matrix terms as $\hat{Y} = \mathbf{X}\beta$.

PRESS: A measure, the smaller the better, of how well the model fits each point in the design. The model is repeatedly refitted to all the design points except the one being predicted. The difference between the predicted value and actual value at each point is then squared and summed over all points to create the PRESS.

Prediction interval: An interval calculated to contain the true mean of a number of individual observations at a stated percentage of the time (usually 95%) in repeated samples.

PV: The square of the standard error of a predicted value.

Prob > F (probability of a larger F-value): The p-value for a test conducted using an F-statistic. If the F-ratio lies in the tail of the F-distribution, the probability of a larger F is small and the variance ratio is judged to be significant. The F-distribution depends on the degrees of freedom (df) for the MS in the numerator and the df of the MS in the denominator of the F-ratio.

Prob > t (probability of a larger t-value): The p-value for a test conducted using a t-statistic. Small values of this probability indicate significance and rejection of the null hypothesis.

Process: Any unit operation, or series of unit operations, with measurable inputs and outputs (responses).

Pure error: Experimental error, or pure error, is the normal variation in the response, which appears when a run is repeated at identical conditions. Repeated runs rarely produce exactly the same results. Pure error is the minimum variation expected in a series of runs. It can be estimated by replicating points in the design. The more replicated points, the better will be the estimate of the pure error.

p-Value: Probability value, usually relating to the risk of falsely rejecting a given hypothesis. Generally, a probability less than 0.05 is considered to be significant.

Quadratic: A second-order polynomial.

Qualitative: See "Categoric variable."

Quantitative: See "Numerical variable."

Randomization: Mixing up runs so that they follow no particular pattern, which is particularly important to ensure that lurking variables do not bias the outcome. Randomization of the order in which experiments are run is essential to satisfy the statistical requirement of the independence of observations.

Range: The difference between the largest and smallest value—a measure of dispersion.

Regression analysis: A method by which data is fitted to a mathematical model.

Regression model: An equation in terms of the coded and actual levels of the variables used to make predictions of a given response, or to describe some physical process of interest.

Replicate: An experimental run performed again from start to finish (not just resampled and/or remeasured). Replication provides an estimate of pure error in the design.

Residual (or "Residual error"): The difference (sometimes referred to as "error") between the observed (actual) response and the value predicted by the model for a particular design point.

Resolution: A measure of the degree of aliasing for the main effects (and interactions) in a fractional factorial design. In general, the resolution of a two-level fractional design is the length of the shortest word in the defining relation. A design of Resolution III confounds the main effects with two-factor interactions. Resolution IV designs confound the main effects with three-factor interactions and two-factor interactions with other two-factor interactions. Resolution V designs confound the main effects with four-factor interactions and two-factor interactions with three-factor interactions.

Response: A measurable product or process characteristic thought to be affected by experimental factors.

Response surface methods: A statistical technique for modeling responses via designed experiments and polynomial equations. The model becomes the basis for 2D contour maps and 3D surface plots for the purposes of optimization.

Risk: The probability of making an error in judgment (i.e., falsely rejecting the null hypothesis). (See also "Significance level.")

Robust: Insensitivity of a process, product, or method to noise factors.

Root MS error: The square root of the residual MS error. It estimates the standard deviation associated with experimental error plus any LOF.

Rotatability: A design property defined by the variance of the predicted response at any point depending only on the distance from the center point. A rotatable design exhibits circular contours on the response surface plot of a standard error for prediction.

RSM: See "Response surface methods."

R-squared: The coefficient of determination. It estimates the fraction (a number between zero and one) of the overall variation in the data accounted for by the model. This statistic indicates the degree of relationship of the response variable to the combined linear predictor variables. Because the raw R-squared statistic is biased, use the adjusted R-squared instead.

R-student: See "Externally studentized residual."

Run: A specified setup of process factors that produces measured response(s) for experimental purposes. (It is also called a "trial.")

Sample: A subset of individuals from a population, usually selected for the purpose of drawing conclusions about specific properties of the entire population.

Saturated: An experimental design with the minimum number of runs required to estimate all effects.

Scheffé polynomial: A form of a mathematical predictive model specifically designed for mixtures. These models are derived from standard polynomials, of varying degrees, by accounting for the mixture constraint that all components sum to a constant. (It is also called a "mixture model.")

Screening: Sifting through a number of variables to find the vital few. Resolution IV two-level fractional factorial designs are often chosen for this purpose.

Significance level: The level of probability, often 0.05, established for rejection of the null hypothesis.

Simplex: A geometric figure with one more vertex than the number of dimensions. For example, the 2D simplex is an equilateral triangle. This shape defines the space for three mixture components and each of them can vary from 0% to 100%.

Standard deviation: A measure of variation in the original units of measure, computed by taking the square root of the variance.

Standard error: The standard deviation of an estimate, for example, the mean, rather than individuals.

Standard error of a parameter: The estimated standard deviation of a parameter or coefficient estimate, a measure of the variability estimate.

Standard order: A conventional ordering of the array of low- and high-factor levels versus runs in a two-level factorial design.

Star points: See "Axial points."

Steepest ascent: The direction one would move from the current point to most quickly increase the response.

Studentized: A value divided by its associated standard error. The resulting quantity is a Z-score (number of standard deviations) useful for purposes of comparison.

Sum of squares (SS): The sum of the squared distances from the mean due to an effect.

Supersaturated: A design matrix having fewer runs than terms in the model.

Synergism: A desirable interaction of two factors where the combination produces a response that is better than what would be expected from either one alone. The same concept can be applied to higher-order interactions.

Term: See "Model coefficient."

Ternary diagram: See "Trilinear graph."

Tolerance interval: A data-based interval constructed by a method that covers a stated proportion of individual observations (typically 99%) at a stated percentage (typically 95%) of the time in repeated samplings.

Transformation: A mathematical conversion of predictor or response values (e.g., $\log(y)$).

Treatment: A procedure applied to an experimental unit. One usually designs experiments to estimate the effect of the procedures (treatments) on the responses of interest.

Trial: See "Run."

Trilinear graph: A scaled (usually from 0% to 100%) equilateral triangle, typically used for the mixture design, with each line from a vertex (apex) to the midpoint on the opposite side of the triangle representing one component. (It is also called a "ternary" graph.)

t-Value: A value associated with the t-distribution that measures the number of standard deviations separating the parameter estimate from zero.

Variable: A factor or response that assumes assigned or measured values.

Variance: A measure of variability computed by summing the squared deviations of individual data from their mean, and then dividing this quantity by the degrees of freedom.

Vertex: A point representing an extreme combination of input variables subject to constraints.

Variance inflation factor (VIF): A measure of how much the standard error for a given model term will be increased due to its correlation with all other terms (R_i^2) in the model. It is defined as

$$VIF = \frac{1}{1 - R_i^2}$$

If a factor is orthogonal to the remaining model terms, its VIF is unity (1). The VIF measures how much the variance of that model coefficient is inflated by the lack of orthogonality in the design. Specifically, the standard error of a model coefficient is increased by a factor equal to the square root of the VIF, when compared to the standard error for the same model coefficient in an orthogonal design. For example, if a coefficient has a VIF of 16, its standard error is four times as large as it would be in an orthogonal design. One or more large VIFs indicate multicollinearity. VIFs exceeding 10 indicate that the associated regression coefficients are poorly estimated due to multicollinearity.

X-matrix: An augmented form of the design matrix with one column for each model coefficient. An (n-by-p) matrix, where n is the number of experiments and p is the number of coefficients including the intercept. The **X**-matrix is used for many of the statistical calculations in the analysis of data sets.

x-Space: See "Design space."

y-Bar ($\bar{y}$): A response mean.

References

Allen, D. M. Mean square error of prediction as a criterion for selecting variables. *Technometrics* 13; 1971: 469–475.

American Society for Metals (ASM), *Handbook of Alloy Phase Diagrams*, ASM International, Ohio, 1992. Available at http://www.amazon.com/ASM-Handbook-Volume-Alloy-Diagrams/dp/0871703815

Anderson, M. *DOE FAQ Alert* (April 2001): Item #3. Minneapolis, Minnesota: Stat-Ease, Inc. (Subscribe to this monthly "e-zine" at www.statease.com/doealertreg.html).

Anderson, M. Six sigma for the road. *Stat-Teaser*. Minneapolis, Minnesota: Stat-Ease, Inc., March 2002: 1–3. http://www.statease.com/news/news0203.pdf.

Anderson, M. and P. Whitcomb. Robust design—Reducing transmitted variation. *The 50th Annual Quality Congress Proceedings*. Milwaukee, Wisconsin: American Society of Quality, 1996, 642–651.

Anderson, M. and P. Whitcomb. How to use graphs to diagnose and deal with bad experimental data. *The 57th Annual Quality Congress Proceedings*: T130. Milwaukee, Wisconsin: American Society of Quality, 2003.

Anderson, M. and P. Whitcomb. Screening process factors in the presence of interactions. *The 58th Annual Quality Congress Proceedings*. Milwaukee, Wisconsin: American Society of Quality, 2004.

Anderson, M. and P. Whitcomb. Employing power to "right-size" design of experiments. *The ITEA Journal* 35; 2014: 40–44.

Anderson, M. and P. Whitcomb. Practical aspects for designing statistically optimal experiments. *Journal of Statistical Science and Application* 2; 2014: 85–92.

Anderson, M. and P. Whitcomb. *DOE Simplified–Practical Tools for Effective Experimentation*, 2nd edn. New York: Productivity Press, 2015.

Anderson, M., Adams, W., and P. Whitcomb. How to properly size response surface method experiment (RSM) designs for system optimization. *The ITEA Journal* 37, no. 1; 2016.

Anantheswaran, R. C. and Y. E. Lin. Studies on popping of popcorn in a microwave oven. *Journal of Food Science* 53, no. 6; 1988: 1746–1749.

Austin, J. G. *Betty Alden: The First-born Daughter of the Pilgrims*, Boston and New York: Houghton, Mifflin and Company, 1891, p. 171.

Bezener, M. Practical Strategies for Model Verification. Webinar, Stat-Ease, Inc., March 2015. Posted at www.statease.com/training/webinar.html.

Block, R. and R. Mee. Table of second-order designs. Via Internet at http://stat.bus .utk.edu/techrpts/2001/2nd_orderdesigns.htm (posted 2001 initially).

Borkowski, J. J. and E. S. Valeroso. Comparison of design optimality criteria of reduced models for response surface designs in the hypercube. *Technometrics* 43; 2001: 468–477.

Box, G. and S. Bisgaard. Design of experiments for discovery, improvement and robustness. (March 1996 short-course, notes taken by MJA.) Center for Quality and Productivity Improvement, University of Wisconsin, Madison, Wisconsin.

Box, G. and N. R. Draper. *Empirical Model-Building and Response Surfaces*. New York: John Wiley and Sons, 1987.

Box, G. and P. Liu. Product design with response surface methods. Report No. 150, Center for Quality and Productivity Improvement, University of Wisconsin, Madison, Wisconsin, May 1998.

Box, G. E. P. Choice of response surface design and alphabetic optimality. *Utilitas Mathematica* 21; 1982: 11–55.

Box, G. E. P. Statistics for discovery. Report No. 179, Center for Quality and Productivity Improvement, University of Wisconsin, Madison, Wisconsin, March 2000.

Box, G. E. P. and D. W. Behnken. Some new three level designs for the study of quantitative variables. *Technometrics* 2; 1960: 455–475.

Box, G. E. P. and D. R. Cox. An analysis of transformations. *Journal of the Royal Statistical Society Series B* 26; 1964: 211.

Box, G. E. P. and J. S. Hunter. Multifactor experimental designs for exploring response surfaces. *The Annals of Mathematical Statistics* 28; 1957: 195–241.

Box, G. E. P. and J. S. Hunter. The 2^{k-p} fractional factorial designs, Part I and II. *Technometrics* 3; 1961: 311–351; 3, no. 4; 1961: 449–458.

Box, G. E. P. and K. B. Wilson. On the experimental attainment of optimum conditions. *Journal of the Royal Statistical Society Series B* 13; 1951: 1–45.

Box, G. J., Hunter, S., and W. Hunter. *Statistics for Experimenters*. New York: John Wiley and Sons, 1978.

Brenneman, W. A. and W. R. Myers. Robust parameter design with categorical noise variables. *Journal of Quality Technology* 35, no. 4; 2003: 335–341.

Bryson, B. A short history of nearly everything, *New York: Broadway Books*, 2003, p. 57.

Burris et al. Trebuchet Report by Team "Hurl." General Engineering project for Professor David Dixon, South Dakota School of Mines and Technology, December 17, 2002.

Chevedden, P. E. et al. The trebuchet. *Scientific American* 273, no. 1; 1995: 66–71.

Cornell, J. *Experiments with Mixtures*, 3rd edn. New York: John Wiley and Sons, 2002.

Crosby, P. *Quality is Free*. New York: McGraw-Hill, 1979.

Derringer, G. C. A balancing act: Optimizing a product's properties. *Quality Progress* 27, no. 6; 1994.

Derringer, G. C. and R. Suich. Simultaneous optimization of several variables. *Journal of Quality Technology* 12; 1980: 214–219.

DeVeaux, R. A guided tour of modern regression methods. In: *1995 Fall Technical Conference: Section on Physical and Engineering Sciences: Proceedings of Conference*, St. Louis, MO, 1995.

DeVeaux, R. *Elements of Experimental Design*. Williamstown, Massachusetts: Williams College, draft version of unpublished book, 2001.

Draper, N. and D. Lin. Small response surface designs. *Technometrics* 32, no. 2; 1990: 187–194.

Draper, N. and H. Smith. *Applied Regression Analysis*. 3rd edn. New York: John Wiley and Sons, 1998.

Draper, N. R. Center points in second-order response surface designs. *Technometrics* 24, no. 2; 1982: 127–133.

Draper, N. R., Davis, T. P., Pozueta, L., and D. M. Grove. Isolation of degrees of freedom for Box–Behnken designs. *Technometrics* 36; 1994: 283–291.

English, T. G. Application of experimental design for efficient wind tunnel testing. Master of Science thesis, Department of Industrial Engineering, Florida State University, Tallahassee, Florida, 2007.

Erbach, T., Fan, L., and S. Kraber. How experimental design optimizes assay automation. *Advance for Medical Laboratory Professionals* 16, no. 12; 2004: 18–21.

Fisher, R. *Statistical Methods for Research Workers*, Oliver and Boyd. Edinburgh, 1925.

Fisher, R. A. Presidential address to the First Indian Statistical Congress. *Sankhya, The Indian Journal of Statistics* 4, no. 1; 1938: 14–17.

Gauss, C. F. Theoria Motus Corporum Celestium. Hamburg, Perthes et Besser, 1809. Translated as Theory of Motion of the Heavenly Bodies Moving about the Sun in Conic Sections (trans. C. H. Davis), Boston, Little, Brown 1857. Reprinted: New York, Dover 1963.

Hadingham, E. Ready, aim, fire! *Smithsonian* 30, no. 10; 2000: 78–87.

Hansen, P. V. Experimental reconstruction of a Medieval Trébuchet by Nyköbing Falster. *Denmark Acta Archaeologica* 63; 1992: 189–268.

Hartley, H. O. Smallest composite designs for quadratic response surfaces. *Biometrics* 15; 1959: 611–624.

Hildebrand, D. and R. L Ott. *Statistical Thinking for Managers*, 4th edn. Belmont, California: Duxbury Press, 1998.

Hosack, H., Marler, N., and D. MacIsaac. Microwave mischief and madness. *The Physics Teacher* 40; 2002: 14–16.

Hunter, J. S. Determination of optimum operating conditions by experimental methods. Parts II-1, 2 & 3. *Industrial Quality Control* 15, no. 6; 1958: 16–24; 15, no. 7; 1959a: 7–15; 15, no. 8; 1959b: 6–14.

Jensen, W. The role of confirmation in designed experiments. *59th Annual Fall Technical Conference of the ASQ and ASA*, October, 2015. Activity Number: 280, Quality and Productivity Section Roundtable Discussion.

Johnson, R.T., Montgomery, D., Jones, B., and Parker, P. Comparing computer experiments for fitting higher-order polynomial metamodels, *Journal of Quality Technology*, 42, no. 1; 2010: 86–102.

Jones, B. and C. Nachtsheim. A class of three-level designs for definitive screening in the presence of second-order effects. *Journal of Quality Technology* 43, no. 1; 2011.

Lin, T. and B. Chananda. Quality improvement of an injection-molded product using design of experiments: A case study. *Quality Engineering* 16, no. 1; 2003: 99–104.

Longley, J. W. An appraisal of least squares programs for the electronic computer from the point of view of the user. *Journal of the American Statistical Association* 62; 1967: 819–841.

Marquardt, D. W. Generalized inverses, ridge regression, biased linear estimation, and nonlinear estimation. *Technometrics* 12; 1970: 591–612.

McKay, M., Beckman, R., and W. Conover. A comparison of three methods for selecting values of input variables in the analysis of output from a computer code. *Technometrics* 21; 1979: 239–245.

Mee, R. New Box–Behnken designs. Unpublished paper documenting current research as of 2003.

Meyer, R. and C. Nachtsheim. The coordinate-exchange algorithm for constructing exact optimal experimental designs. *Technometrics* 37; 1995: 60–69.

Montgomery, D. *Design and Analysis of Experiments*, 8th edn. New York: John Wiley and Sons, 2012.

Montgomery, D., Peck, E., and G. Vining. *Introduction to Linear Regression Analysis*, 5th edn. New York: John Wiley and Sons, 2012.

Myers, R. *Classical and Modern Regression with Applications*. Boston: Duxbury Press, 1986.

Myers, R. and D. Montgomery. *Response Surface Methodology*, 2nd edn. New York: John Wiley and Sons, 2002.

Myers, R., Montgomery, D., and C. Anderson-Cook. *Response Surface Methodology*, 4th edn. New York: John Wiley and Sons, 2016.

Nelder, J. A. The selection of terms in response-surface models—How strong is the weak-heredity principle. *The American Statistician* 52, no. 4; 1998: 315–318.

Nelder, J. A. and R. Mead. A simplex method for function minimisation. *Computational Journal* 7; 1965: 308–313.

Newman, J. R. (ed.). *The World of Mathematics*, New York: Simon & Schuster, 1956.

Oehlert, G. *A First Course in Design and Analysis of Experiments*. New York: W. H. Freeman, 2000.

Oehlert, G. and P. Whitcomb. Small, efficient, equireplicated resolution V fractions of 2^k designs and their application to central composite designs. *46th Annual Fall Technical Conference of the ASQ and ASA*, Valley Forge, PA, October 2002.

Peixoto, J. L. A property of well-formulated polynomial regression models, *The American Statistician* 44, no. 1; 1990: 26–30.

Peterson, I. Super Bowls and stock markets. *Science News* 157, 2000: 399. Available at http://www.sciencenews.org/article/super-bowls-and-stock-markets-0

Peterson, I. Completing Latin squares. *Science News Online* 157, no. 19; Week of May 6, 2000. https://www.sciencenews.org/article/completing-latin-squares.

Pyzdek, T. *The Six Sigma Handbook*, 2nd edn. New York: McGraw-Hill, 2003.

Richert, S. H., Morr, C. V., and C. M. Cooney. Effect of heat and other factors upon foaming properties of whey protein concentrates. *Journal of Food Science* 39; 1974: 42–48.

Roquemore, K. G. Hybrid designs for quadratic response surfaces. *Technometrics* 18; 1976: 419–423.

Rossman, A. Televisions, physicians, and life expectancy. *Journal of Statistics Education* 2, no. 2; 1994. http://www.amstat.org/publications/jse/v2n2/datasets.rossman.html.

Sachar, L. *Holes.* New York: Yearling Books, 201.

Safire, W. On Language. *The New York Times Magazine*, April 27, 2003, p. 22.

Santner, T., Williams, B., and W. Notz. *The Design and Analysis of Computer Experiments.* New York: Springer-Verlag, 2003.

Scheffé, H. Experiments with mixtures. *Journal of the Royal Statistical Society. Series B: Methodological* 20; 1958.

Sergent, M. et al. Correct and incorrect use of multilinear regression. *Chemometrics and Intelligent Laboratory Systems* 27; 1995: 153–162.

Sies, H. The storks and the babies. *Nature* 322; 1988: 495.

Smith, R. et al. *Physical Science.* Westerville, Ohio: Glencoe Div. of MacMillan/McGraw-Hill, 1993.

Smith, W. F. Jr. *Experimental Design for Formulation.* American Statistical Association (ASA), Alexandria, VA and the Society for Industrial and Applied Mathematics (SIAM), Philadelphia, PA, 2005.

Snee, R. Validation of regression models: Methods and examples. *Technometrics* 10; 1977: 415–428.

Snipes, M. and D. C. Taylor. Model selection and Akaike information criteria: An example from wine ratings and prices. *Wine Economics and Policy* 3, no. 1; 2014: 3–9.

Spendley, W., Hext, G. R., and F. R Himsworth. Sequential application of simplex designs in optimisation and evolutionary operation. *Technometrics* 4, no. 4; 1962: 441–461.

Stanard, C. L. *Multiple Response Optimizations Using Designed Experiments: A Practical Example.* Technical Information Series, 2002GRC120, GE Global Research, Schenectady, NY, May 2002.

Stigler, S. *The History of Statistics.* Cambridge: Belknap Press of Harvard University Press, 1986.

Sullivan, E C. and W C. Taylor. Glass. United States Patent No. 1,304,623 assigned to Corning Glass Works, Corning, New York, May 27, 1919.

Taylor, W. *Comparing Three Approaches to Robust Design: Taguchi versus Dual Response Surface versus Tolerance Analysis.* Round Lake, Illinois: Baxter Healthcare Corporation, 1996.

Thomas, G. et al. *Thomas' Calculus*, 10th edn. Boston, Massachusetts: Addison-Wesley Publishing, 2000.

Unal, R., Lepsch, R. A., Jr., and M. L. McMillin. Response surface model building and multidisciplinary optimization using overdetermined D-optimal designs. *7th AIAA/USAF/NASA/ISSMO Symposium on Multidisciplinary Analysis and Optimization*, St. Louis, Missouri, September 2–4, 1998: 405–411.

Unal, R., Wu, K. C., and D. O. Stanley. Structural design optimization for a space truss platform using response surface methods. *Quality Engineering* 9, no. 3; 1997: 441–447.

US Department of Defense Federal Acquisition Regulation 246.101. Available at http://www.acq.osd.mil/dpap/dars/dfars/pdf/r20031001/246_1.pdf

Veal, E. F. and A. Mackey. Bread quality as affected by flour–water ratio, temperature and speed of mixing during early stages of dough preparation. Home Economics Research, Technical Paper 859, Oregon Agricultural Experiment Station, Oregon State University, Corvallis, Oregon, 2000.

Vining, G. Technical advice: Residual plots to check assumptions. *Quality Engineering*, 23, no. 1; 2011: 105–110.

Vining, G. G., Kowalski, S. M., and Montgomery, D. C. Response surface designs within a split-plot structure. *Journal of Quality Technology*, 37, no. 2; 2005: 115–129.

Weisberg, S. *Applied Linear Regression*, 4th edn. New York: John Wiley and Sons, 2013.

Westlake, W. J. Composite designs based on irregular fractions of factorials. *Biometrics* 21; 1965: 324–336.

Whitcomb, P. and M. Anderson. Practical versus statistical aspects of altering central composite designs. Activity 141, Roundtable discussion at *Joint Statistical Meetings of the American Statistical Association*. San Francisco, California, August 4, 2003. (Notes available from authors.)

Zahran, A. R., Anderson-Cook, C., and R. Myers. Fraction of design space to assess prediction capability of response surface designs. *Journal of Quality Technology* 35; 2003: 377–386.

Zink, P. S. et al. Impact of active aeroelastic wing technology on wing geometry using response surface methodology. Talk at *Langley International Forum on Aeroelasticity and Structural Dynamics*, Williamsburg, Virginia, June 22–25, 1999.

Index

About the Software

To make RSM easy, this book is augmented with a fully functional, time limited version of a commercially available computer program from Stat-Ease, Inc. called Design-Expert software. Download this Windows-based computational tool, as well as companion files in Adobe's portable document format that provide tutorials on various designs from the very basics to sophisticated RSM, from www.statease.com/rsm-simplified-2.html. There, you will also find files of data for most of the exercises in the book. The datasets are named to be easily cross-referenced with the corresponding material in the book. Also, if you work through all the problems (the only way to a working knowledge of RSM), check your answers by downloading them from the book's website.

You are encouraged to reproduce the results shown in the book and to explore them further. The Stat-Ease software offers far more detail in statistical outputs and many more graphics than can be included in this book. You will find a great deal of information on program features and statistical background in the online hypertext help system built into the software.

Technical support for the software can be obtained by contacting:

Stat-Ease, Inc.
2021 East Hennepin Ave, Suite 480
Minneapolis, Minnesota 55413
Telephone: 612-378-9449
Fax: 612-378-2152
E-mail: support@statease.com
Website: www.statease.com

Printed and bound by CPI Group (UK) Ltd, Croydon, CR0 4YY

23/10/2024

01777692-0011